EXAMPLES

IN

FINITE DIFFERENCES, CALCULUS
AND PROBABILITY

EXAMPLES

IN

FINITE DIFFERENCES, CALCULUS AND PROBABILITY

SUPPLEMENT TO
*AN ELEMENTARY TREATISE ON
ACTUARIAL MATHEMATICS*

By

HARRY FREEMAN, M.A., F.I.A.

CAMBRIDGE

Published for the Institute of Actuaries

AT THE UNIVERSITY PRESS

1936

CAMBRIDGE
UNIVERSITY PRESS

University Printing House, Cambridge CB2 8BS, United Kingdom

Cambridge University Press is part of the University of Cambridge.

It furthers the University's mission by disseminating knowledge in the pursuit of education, learning and research at the highest international levels of excellence.

www.cambridge.org
Information on this title: www.cambridge.org/9781316612781

© Cambridge University Press 1936

First published 1936
First paperback edition 2016

A catalogue record for this publication is available from the British Library

ISBN 978-1-316-61278-1 Paperback

CONTENTS

PREFACE

S INCE the publication in 1931 of *An Elementary Treatise on Actuarial Mathematics* it has been suggested on many occasions that the usefulness of the book would be increased by the addition of further examples. The inclusion of additional examples in the book as it stood was impracticable, and it appeared that the difficulty could only be overcome by the publication of a supplement to the book. This supplement accordingly contains a further selection of problems on Finite Differences, Calculus and Probability, together with some hints for solution. It has been considered neither necessary nor advisable to include a hint for the solution of every question, and, moreover, in only a few instances has the complete solution of the question been given.

It is hoped that the supplement will prove of value to students, especially to those who have completed the course for the examination and who desire to attempt the solution of extra problems for the purpose of revision.

H. F.

April 1936

FINITE DIFFERENCES

Unless otherwise stated, the interval of differencing is to be taken as unity.

1. Find $\Delta^2 (2^{4x} - x^2 + x - 1)$.

2. Express in its simplest form $\dfrac{\Delta^3}{E^2} u_x - \dfrac{\Delta^3 u_x}{E^2 u_x}$, where

   (i) $u_x = a + bx^3$; (ii) $u_x = e^{hx}$ (interval h).

3. Obtain, for an interval α,

   (i) $\Delta \sin x \cos x$; (ii) $\dfrac{1}{\Delta \cot x} - \dfrac{1}{\Delta \tan x}$.

4. Find $\Delta^3 \dfrac{4x + 17}{(2x + 1)(2x + 3)(2x + 5)(2x + 7)}$.

5. Obtain simple expressions for the first differences of

   (i) $(4 - x)\{(x - 2)^2 - \frac{1}{4}\}^{-1}$;

   (ii) $[\Delta E^{-1} \log_e (ax + b)]^2$ (interval a).

6. Find the value of
$$\Delta x^{(m)} - 2\Delta^2 x^{(m)} + 3\Delta^3 x^{(m)} - 4\Delta^4 x^{(m)} + \ldots \text{ to } m \text{ terms.}$$

7. Calculate $\Delta E^{-2} f(x)$, when

   (i) $f(x) = xe^x$; (ii) $f(x) = (3 + x)[(x + 2)^2 - \frac{1}{9}]^{-1}$.

8. If $u_x = e^x$ and $v_x = x^3$, find the value of

   (i) $\dfrac{E u_x}{\Delta v_x} - \dfrac{\Delta^2}{E^2} u_x + \dfrac{\Delta^2}{E^2} v_x$; (ii) $\Delta (u_x v_x) - u_x \Delta v_x - v_{x+1} \Delta u_x$.

9. Show that
$$u_0 + nu_1 x + \frac{n(n-1)}{2!} u_2 x^2 + \ldots$$
$$= (1 + x)^n u_0 + n(1 + x)^{n-1} x\Delta u_0 + \frac{n(n-1)}{2!} (1 + x)^{n-2} x^2 \Delta^2 u_0 + \ldots.$$

10. (i) Given $u_0 = 5$, $u_1 = 10$, $u_2 = 20$, $u_3 = -8$, obtain u_4 by constructing a difference table.

(ii) $u_x = \cdot 1126$, $u_{x+1} = \cdot 1365$, $u_{x+3} = \cdot 2143$. Find u_{x+2}.

11. Without constructing a difference table find $\Delta^6 u_3$, given:

x	9	8	7	6	5	4	3
u_x	57	54	41	26	13	4	0

12. Corresponding values of x and $f(x)$ are as in the following table:

x	10	11	12	13	14	15	16
$f(x)$	0	10	α	51	100	200	β

If fifth differences are constant and equal to 100, find α and β.

13. $u_0 = 0$, $u_1 = 79$, $u_2 = 146$, $u_3 = 204$, $u_5 = 310$, $u_8 = 534$. Find u_4, u_6 and u_7.

14. Given that as x takes the values 0, 1, 2, 3, 4, 5, u_x takes the values 9, 1, 25, 81, 169, 289 respectively, obtain the form of u_x in the following cases:

(a) u_x is a rational integral function of x of the second degree;

(b) u_x is a rational integral function of x of degree less than 6;

(c) u_x is a rational integral function of x;

(d) no information is available regarding the nature of u_x.

15. $q_{55} = \cdot 02111$, $q_{57} = \cdot 02444$, $q_{58} = \cdot 02629$, $q_{59} = \cdot 02827$. Find q_{56}.

16. Supply the annuity values for ages 40 and 43:

Age x	38	39	41	42	44
a_x	13·798	13·408	12·605	12·194	11·355

17. The following values of x and $f(x)$ are given:

x	0	1	2	3	4	5
$f(x)$	−66	0	0	0	354	1824

Assuming that $f(x)$ is a rational integral function of x, find the form of $f(x)$, and hence obtain $f(\frac{1}{4})$ to the nearest whole number.

18. $u_{-2} = \cdot 1126$, $u_{-1} = \cdot 1365$, $u_1 = \cdot 1849$, $u_3 = \cdot 2381$. Find u_0 and u_2.

19. Find A_{51}, given that $A_{43} = \cdot 49136$, $A_{44} = \cdot 50515$, $A_{45} = \cdot 51851$, $A_{46} = \cdot 53148$, $A_{47} = \cdot 54410$.

20. Complete the series for $f(x)$ from $x = 0$ to $x = 12$ from the data:

x	0	4	8	12
$f(x)$	2·714	2·884	3·037	3·175

21. Express $\dfrac{x^3+x+5}{x^4+10x^3+35x^2+50x+24}$ as the sum of a number of inverse factorials, and hence find its second difference in the form of a single fraction.

22. Six consecutive terms of a series are given:

$$0,\ 1,\ 3,\ 8,\ 20,\ 47.$$

An estimate is required of the seventh term, and A gives 103 for his answer. B states that a more probable figure is 105. Examine these estimates and give reasons for your preference, if any, for one over the other.

23. The following table gives the number of children under five years of age per 1000 women between the ages of 16 and 44 resident in a certain country at the undermentioned dates:

Year	1800	1805	1815	1830	1860
Number of children	976	976	952	877	714

Estimate the number of children per 1000 women for the year 1810.

24. Find u_x as a rational integral function of x from the following data:

x	2	6	4	8	7
u_x	5	205	57	497	330

25. What simple form of function gives $u_0 = -4$, $u_3 = 23$, $u_4 = 60$ and $u_{10} = 996$?

26. Find x when $u_x = 160$, given that u_x has the values 153, 157, 164 and 177 when x has the values 10, 11, 13 and 16 respectively.

27. Four values of a function u_a, u_x, u_b, u_c are given, corresponding to the values a, x, b, c of the variable. Find the third divided difference of u_a and deduce Lagrange's formula for u_x in terms of u_a, u_b and u_c.

28. Prove the interpolation formula:

$$u_x = u_0 + x\cdot\tfrac{1}{2}(\Delta u_0 + \Delta u_1) + \frac{x^2}{2}\Delta^2 u_{-1}$$
$$+ \frac{x(x^2-1)}{6}\cdot\tfrac{1}{2}(\Delta^3 u_{-1} + \Delta^3 u_{-2}) + \frac{x^2(x^2-1)}{24}\Delta^4 u_{-2} + \ldots.$$

Set out a table of differences and find q_{51} from the values

$$q_{50} = \cdot015714,\quad q_{48} = \cdot015152,\quad q_{52} = \cdot016303,$$
$$q_{54} = \cdot016915,\quad q_{46} = \cdot014625.$$

29. Find $f(1\cdot4375)$, given the following data:

x	$1\cdot40$	$1\cdot41$	$1\cdot42$	$1\cdot43$	$1\cdot44$	$1\cdot45$	$1\cdot46$
$f(x)$	1053	808	608	452	339	268	238

30. Given n corresponding values of a variable and of a rational integral function of the variable, in what circumstances would you use the following formulae to interpolate between two given values of the function to find a value corresponding to an intermediate value of the variable:

(i) Newton's advancing difference formula;

(ii) Lagrange's formula;

(iii) Newton's divided difference formula;

(iv) A central difference formula?

31. Use an appropriate interpolation formula to find u_1, given $u_x = 0$, 108, 200, 316, 508, 840 for the values of $x = -4, -2, 0, 2, 4, 6$ respectively.

32. Show that

$$u_x = \tfrac12(u_0+u_1)+(x-\tfrac12)\,\Delta u_0 + \frac{x(x-1)}{2!}\cdot\frac{\Delta^2 u_{-1}+\Delta^2 u_0}{2}+\dots$$

Extend the formula to include terms involving the third and fourth differences of the function.

Use the formula to find $f(50)$ from the values:

x	40	48	56	64
$f(x)$	$14\cdot27$	$15\cdot81$	$17\cdot72$	$19\cdot96$

33. From the following data calculate a_{26}:

x	15	20	25	30	35	40
a_x	$24\cdot584$	$23\cdot473$	$22\cdot285$	$20\cdot943$	$19\cdot468$	$17\cdot881$

using (i) two of the values;

(ii) four of the values;

(iii) all the values given.

34. Construct a formula of interpolation expressing u_x in terms of u_0, Δu_{-1}, $\Delta^2 u_{-1}$, $\Delta^3 u_{-2}$, $\Delta^4 u_{-2}$, $\Delta^5 u_{-3}$ and $\Delta^6 u_{-3}$, and hence determine the entry corresponding to the argument $4\tfrac12$, given:

Argument	1	2	3	4	5	6	7
Entry	1901	1264	29	-304	5125	27056	85349

35. Interpolate by means of Gauss's forward formula to find the present value of 1 due 27 years hence at 5 per cent. compound interest from the following data:

No. of years	15	20	25	30	35	40
Present value	10·3797	12·4622	14·0939	15·3725	16·3742	17·1591

36. Employ an appropriate formula to obtain successive approximations to $f(28·3)$, given the table:

x	26	27	28	29	30
$f(x)$	·038462	·037037	·035714	·034483	·033333

37. Prove Everett's formula:

$$u_x = xu_1 + \frac{x(x^2-1)}{3!}\Delta^2 u_0 + \frac{x(x^2-1)(x^2-4)}{5!}\Delta^4 u_{-1} + \dots$$

$$+ \xi u_0 + \frac{\xi(\xi^2-1)}{3!}\Delta^2 u_{-1} + \frac{\xi(\xi^2-1)(\xi^2-4)}{5!}\Delta^4 u_{-2} + \dots.$$

What are the practical advantages of this formula?

Note. In the next five questions Everett's formula is to be used.

38. Find the present value of £1 per annum at the end of 20 years at 3·2 per cent., given the following extract from tables of compound interest:

Rate per cent.	2	2½	3	3½	4	4½
Present value of 1 per annum (£)	16·3514	15·5892	14·8775	14·2124	13·5903	13·0079

39. The annual premium (P_x) for an assurance of 1 at 4 per cent. being available for the undermentioned ages, find P_x for ages 36 to 44 inclusive:

Age x	25	30	35	40	45	50	55
P_x	·01431	·01662	·01961	·02351	·02865	·03550	·04471

40. The following is an extract from the prospectus of a Life Assurance Company, giving the annual premiums for an assurance of £100:

Age at entry	Annual premium		
	£	s.	d.
20	1	19	6
25	2	3	11
30	2	9	5
35	2	16	2
40	3	4	10
45	3	15	8
50	4	10	7

Obtain the premiums for ages 31 to 39 inclusive.

41. Calculate q_{71}, q_{72}, q_{73} and q_{74}, given:

Age x	60	65	70	75	80	85
q_x	·03108	·04468	·06558	·09747	·14525	·21504

42. From the following table of tan x obtain the values of the tangents of all angles from 55° to 65° inclusive:

Angle x (degrees)	45	50	55	60	65	70	75
tan x	1·0000	1·1918	1·4281	1·7320	2·1445	2·7475	3·7320

43. Values of the yearly pension secured to a wife after her husband's death by a yearly contribution of £1 are given for a husband aged 30 next birthday, according to two different tables:

Age of wife next birthday	20	30	40	50	60
Pension on Table I (£)	4·39	5·00	6·03	7·77	10·74
Pension on Table II (£)	3·90	4·47	5·44	7·13	10·15

Find, for age of wife 33 next birthday, the percentage increase in the yearly pension resulting from the adoption of Table I in place of Table II.

44. Use Stirling's formula to find a_{32}, given:

Age x	20	25	30	35	40	45
a_x	14·035	13·674	13·257	12·734	12·089	11·309

How would you alter your method, if at all, if you were required to find the values for all ages from 20 to 45 inclusive?

45. If a_x is the value of an annuity on a life aged x, payable yearly, an approximation to the value of the annuity payable continuously—i.e. at very frequent intervals—is

$$\bar{a}_x = \tfrac{1}{2} + a_x - \tfrac{1}{12}\,(\mu_x + \delta).$$

a_x is available for the undermentioned ages:

Age x	75	80	85	90	95
a_x	4·747	3·417	2·419	1·676	1·045

and the corrective factor for a different set of ages, thus:

Age x	77	82	87	92
$\tfrac{1}{12}(\mu_x + \delta)$	·013	·019	·027	·039

From these data calculate $\bar{a}_{83}$.

46. $f(8) = 1·4422$, $f(9) = 1·4581$, $f(10) = 1·4736$, $f(11) = 1·4888$ and $f(12) = 1·5037$. Find $f(8·5)$, using all the values given.

47. Use a central difference formula to obtain successive approximations to $(1 + i)^{17}$ from the table:

x	0	5	10	15	20	25	30
$(1+i)^x$	1	1·27628	1·62889	2·07893	2·65330	3·38635	4·32194

48. From the following values of the expectation of life by English Life Table No. 8, estimate e_{28}:

x	20	25	30	35
e_x	44·21	40·00	35·81	31·71

If in addition the values for ages 10 (53·08) and 15 (48·57) are given, obtain a revised estimate for e_{28}.

49. Construct a table of divided differences from the data:

x	0	5	8	9	25
$f(x)$	−179	1	13	37	7221

Extend the table to include arguments $x = 5$, repeated as many times as may be necessary, and thus find $f(x)$ in powers of $(x-5)$.

50. Prove Everett's second formula:

$$u_{p-\frac{1}{2}} = u_0 + \frac{p^2 - \tfrac{1}{4}}{2!}\,\Delta u_0 + \frac{(p^2 - \tfrac{1}{4})\,(p^2 - \tfrac{9}{4})}{4!}\,\Delta^3 u_{-1} + \dots$$

$$- \frac{q^2 - \tfrac{1}{4}}{2!}\,\Delta u_{-1} - \frac{(q^2 - \tfrac{1}{4})\,(q^2 - \tfrac{9}{4})}{4!}\,\Delta^3 u_{-2} - \dots$$

When would you use this formula in practice?

51. Comment on the following extracts from a text-book on statistics:

(a) In order to apply the method of interpolation to statistics, it is assumed that the formulae may be employed with respect to figures which do not obey any definite law.

(b) When the argument proceeds by unequal intervals, the most convenient formula for interpolation is Lagrange's.

52. Establish the formula:

$$u_x = -\tfrac{1}{6}\left(x^2 - \tfrac{1}{4}\right)\left(x - \tfrac{3}{2}\right)u_{-\frac{3}{2}} + \tfrac{1}{2}\left(x - \tfrac{1}{2}\right)\left(x^2 - \tfrac{9}{4}\right)u_{-\frac{1}{2}}$$
$$ - \tfrac{1}{2}\left(x + \tfrac{1}{2}\right)\left(x^2 - \tfrac{9}{4}\right)u_{\frac{1}{2}} + \tfrac{1}{6}\left(x^2 - \tfrac{1}{4}\right)\left(x + \tfrac{3}{2}\right)u_{\frac{3}{2}}.$$

Use the formula to find $f\left(1\tfrac{1}{4}\right)$, given that $f(0) = 896$, $f(1) = 1000$, $f(2) = 1216$, $f(3) = 1508$.

53. Given $f\left(-\tfrac{5}{2}\right)$, $f\left(-\tfrac{3}{2}\right)$, $f\left(-\tfrac{1}{2}\right)$, $f\left(\tfrac{1}{2}\right)$, $f\left(\tfrac{3}{2}\right)$ and $f\left(\tfrac{5}{2}\right)$, obtain a six-term formula for $f(x)$ similar to the four-term formula in Qu. 52.

54. In a certain series $u_0 = a$, $\Delta u_0 = b$, $\Delta^2 u_x = c$ for all values of x from 0 to $m-1$, and $\Delta^2 u_x = d$ for all subsequent values; find u_{m+n}, where m and n are positive.

If the series in question is

find u.   6, 11, 20, 33, 50, u, 105, 143, 188, 240, ...,

55. y is a function of x. When x is 30, y is 1; when x is 31, y is 2; and when x is 32, y is 103. Find a value or values for x when y is 50, illustrating your answer by a diagram.

56. The following table gives the present value of £1 due 20 years hence, at varying rates of interest. Find, correct to three decimal places, the rate of interest for which the present value is £·4, by a method of inverse interpolation:

Rate per cent.	$4\tfrac{3}{8}$	$4\tfrac{1}{2}$	$4\tfrac{5}{8}$	$4\tfrac{3}{4}$	$4\tfrac{7}{8}$	5
Present value of £1	·42469	·41464	·40485	·39529	·38598	·37689

57. Find approximate values of the real roots of the equations

(a) $x^3 + x = 3$ (root between 1·2 and 1·3);

(b) $x^3 + 2x = 20$ (root between 2·4 and 2·5).

58. If $u_x = p + x + x^2$, find, by inverse interpolation from u_0, u_1 and u_4, the value of x corresponding to $u_x = 3$.

What must p be in order that this value of x may satisfy the equation

$$3 = p + x + x^2?$$

59. Explain the method of elimination of third differences as used in inverse interpolation, and employ the method to find the real root of $x^3 + 12x = 12$, correct to three decimal places.

60. One root of the equation $10(x^3 - 7x + 6) = x^2$ is approximately 1. Find its value correct to four decimal places.

61. The value of an annuity-certain of £1 per annum for 30 years is given as £20, and it is desired to find the rate of interest involved. Using the following extract from a table of annuity values, obtain an approximate value for the rate of interest:

Rate per cent.	2	$2\frac{1}{2}$	3	$3\frac{1}{2}$
Value of a 30-year annuity (£)	22·40	20·93	19·60	18·39

62. Find the real roots of

(a) $x^3 = 6x + 11$;   (b) $13x^3 + 68x = 57x^2 + 36$;

each correct to three decimal places.

63. u is the present value of a given series of payments, and it is known that u lies between u_0 and u_1, which are the present values of the given series at rates i and $i + h$ respectively. If $i + k$ be the rate of interest for which the present value is u, prove that, approximately,

$$k = \frac{h\left(1 + \frac{1}{2}\Delta^2 u_0/\Delta u_0\right)}{\{\Delta u_0/(u - u_0)\} + \{\frac{1}{2}\Delta^2 u_0/\Delta u_0\}}.$$

64. If $\alpha = \Delta u_0$ and $\beta = \frac{1}{2}\Delta^2(u_{-1} + u_0)$, show that the relation between k and h in Qu. 63 may be expressed as

$$k = h \frac{u - u_0}{\alpha - \frac{1}{2}\beta + \frac{1}{2}\beta \frac{u - u_0}{\alpha - \frac{1}{2}\beta}} \quad \text{approximately.}$$

65. Show, by means of examples, that the following are not universally true:

(i) $\Sigma\Delta \equiv 1$;   (ii) $u_x = (1 + \Delta)^x u_0$.

66. Sum to n terms

(a) 3, 8, 11, 18, 35, 68, ...;   (b) 0, 0, $1\frac{1}{2}$, 5, $10\frac{3}{4}$, $18\frac{7}{8}$, $29\frac{7}{16}$, ...

67. Find the sum of n terms of the series whose xth term is

$$3^x (3x - 3)(3x - 6) - 2x (2x - 1)(2x - 2).$$

68. Sum to n terms

      (a) 2, 12, 22, 36, 62, 116, 230, ...;

      (b) 1, 2, 4, 14, 40, 92, 184, 338, ...

69. Prove the formula for summation by parts in finite differences, and apply it to obtain the sum of n terms of the series

$$5 + 2.8 + 4.12 + 8.17 + 16.23 + ...$$

70. Sum to n terms

      (a) $\dfrac{3}{1.2.4} + \dfrac{4}{2.3.5} + \dfrac{5}{3.4.6} + \dfrac{6}{4.5.7} + ...$;

      (b) $\dfrac{15}{1.2.3}\left(\dfrac{6}{7}\right) + \dfrac{16}{2.3.4}\left(\dfrac{6}{7}\right)^2 + \dfrac{17}{3.4.5}\left(\dfrac{6}{7}\right)^3 + ...$

71. Find the general term of each of the series

      (a) 67, 39, 29, 29, 35, 47, 69, ...;

      (b) 67, 39, 29, 29, 35, 45, 58, ...

Sum series (b) to n terms.

72. The series 10, 30, 98, 346, 1278, 4838, ... is the sum of two series. The xth term of one of the series is of the form $p + qx + r^x$, while the other series is a geometrical progression. Find the general term of the combined series and its sum to n terms.

73. Sum to n terms

$$1.1 + 32.32 + 64.64 + 97.98 + 131.136 + 166.182 + ...$$

74. Find

      (i) $\Delta^{-1}\{x! (x^2 + x + 1)\}$;   (ii) $\Delta^{-n}3^x - \dfrac{1}{\Delta^n 2^x}$.

75. Evaluate $\overset{n}{\underset{1}{\Sigma}} \{\Sigma x^2 a^x\}$.

76. Prove that $2\overset{n}{\underset{1}{\Sigma}} \{x 2^x + 3x\} = (n - 1) 2^{n+2} + 3n^2 + 3n + 4.$

77. $\Delta^3 u_x = 2^x (x^2 + 12x + 27)$. Find the form of u_x and the value of $\overset{6}{\underset{1}{\Sigma}} u_x$, given that $u_{-1} = -1$, $u_0 = -3$, $u_1 = -4$.

78. Sum to infinity

$$\frac{1}{1!} + \frac{1}{2!} + \frac{2}{3!} + \frac{10}{4!} + \frac{31}{5!} + \frac{71}{6!} + \dots.$$

79. The xth term of the series 2, 16, 68, 244, ... is of the form $ab^x + cd^x$. Sum the series to n terms.

80. Evaluate $\Delta^{-1} [2^x (3x^3 - 4x^2 + x - 9)]$.

81. u_x is the xth term of the series

$$2, \ 6, \ 11, \ 20, \ 36, \ 62, \ \dots;$$

$v_x = \overset{x}{\underset{1}{\Sigma}} u_x$. Express $\overset{n}{\underset{1}{\Sigma}} v_x$ in an ascending series of powers of n.

82. Assuming that $\Delta^3 u_x$ is constant, find u_5, given that $u_1 = 8$, $u_2 + u_3 = 26$, $u_4 + u_5 + u_6 = 139$, $u_7 + u_8 + u_9 + u_{10} = 1002$.

83. Find u_1, given

$$u_{10} = 235, \quad \overset{10}{\underset{1}{\Sigma}} u_x = 715, \quad \overset{10}{\underset{4}{\Sigma}} u_x = 693 \quad \text{and} \quad \overset{10}{\underset{7}{\Sigma}} u_x = 596.$$

84. Given that the sum of the first five terms of a series is 1365, the sum of the next five 3790, the sum of the next five 8215 and the sum of the next five 15265, find the form of the rational integral function of the third degree which represents the general term of the series.

85. Sum to n terms:

(i) $3^2.2.1 + 4^2.3.2 + 5^2.4.3 + 6^2.5.4 + \dots;$

(ii) $16 + 64.22 + 256.79 + 1024.190 + 4096.373 + \dots.$

86. Prove that, to third differences,

$$u_7 = \cdot 2 v_5 - \cdot 008 (v_{10} - 2 v_5 + v_0),$$

where $v_k = u_k + u_{k+1} + u_{k+2} + u_{k+3} + u_{k+4}.$

Find u_7, u_{12} and u_{17}, given:

k	0	5	10	15	20
v_k	21·35	73·35	122·95	170·40	215·85

87. A series consists of nine terms. The sum of the first and ninth terms of the series is 73·84; of the second and eighth 73·07; of the third and seventh 72·52; and of the fourth and sixth 72·19. Find the middle term of the series.

88. The series -31, -28, -25, -19, -1, 56, ... can be split up into two series, one a rational integral function of x, and the other a geometrical progression. Find the sum of the first ten terms of that part of the series represented by the rational integral function.

89. Prove that if the nth differences of a function are in geometrical progression, the differences of the next higher order form a geometrical progression with the same common ratio.

If $u_0 = 1$ and $\Delta u_x = 3u_x + 2$ for all positive integral values of x, find the form of u_x.

90. If u_x is a rational integral function of the third degree and $v_x = u_{x-1} + u_x + u_{x+1}$ for all integral values of x, express u_x in terms of v_{x-1}, v_x and v_{x+1}.

91. Prove that, if a_n diminishes as n increases and converges to the limit zero, the sum to infinity of the series $a_1 - a_2 + a_3 - \ldots$ is the same as the sum to infinity of the series $\frac{1}{2}a_1 - \frac{1}{4}\Delta a_1 + \frac{1}{8}\Delta^2 a_1 - \ldots$.

Find the sum to infinity of the series $\frac{1}{10} - \frac{1}{11} + \frac{1}{12} - \ldots$ true to four decimal places.

92. Assuming that

$$\frac{(x+y)^{(n)}}{n!} = a_0 + a_1 x + a_2 \frac{x^{(2)}}{2!} + a_3 \frac{x^{(3)}}{3!} + \ldots + a_n \frac{x^{(n)}}{n!},$$

where a_0, a_1, a_2, ..., a_n are functions of y only, show that, by differencing both sides r times and putting $x = 0$, a_r can be found.

Hence deduce Vandermonde's Theorem.

93. Prove that

$$u_1 \frac{x}{1} - u_2 \frac{x^2}{2} + u_3 \frac{x^3}{3} - \ldots$$

$$= u_0 \log_e (1+x) + \frac{x}{1+x} \Delta u_0 - \frac{x^2}{2(1+x)^2} \Delta^2 u_0 + \frac{x^3}{3(1+x)^3} \Delta^3 u_0 - \ldots.$$

94. Use the method of separation of symbols to find the value of
$$c_0 a - c_1 (a-1) + c_2 (a-2) - \dots,$$
where c_r is the coefficient of x^r in the expansion of $(1+x)^n$ in ascending powers of x, and n is a positive integer.

Note. In the next four questions $c_0, c_1, c_2, \dots, c_n$ have the same meanings as in Qu. 94.

95. Find the value of $\dfrac{c_0}{x} - \dfrac{c_1}{x+1} + \dfrac{c_2}{x+2} - \dots + (-1)^n \dfrac{c_n}{x+n}$.

96. Prove that
$$(n-1)^2 c_1 + (n-3)^2 c_3 + (n-5)^2 c_5 + \dots = 2^{n-3} n(n+1).$$

97. If n is a positive integer greater than 3, prove that
$$n^3 + c_2 (n-2)^3 + c_4 (n-4)^3 + \dots = n^2 (n+3) 2^{n-4}.$$

98. Find the value of $c_0 n^n - c_1 (n-2)^n + c_2 (n-4)^n - \dots.$

99. If r be an integer greater than unity, find the sum of the series
$$ab - (r+1)(a-1)(b-1) + \frac{(r+1)r}{2!}(a-2)(b-2)$$
$$- \frac{(r+1)r(r-1)}{3!}(a-3)(b-3) + \dots.$$

100. Find the sum to infinity of the series
$$x^2 + \tfrac{1}{2}(x+1)^2 + (\tfrac{1}{2})^2 (x+2)^2 + (\tfrac{1}{2})^3 (x+3)^2 + \dots.$$

101. Sum the series
$$\tfrac{1}{2} - \tfrac{1}{3}m + \tfrac{1}{4}m(m-1)/2! - \tfrac{1}{5}m(m-1)(m-2)/3! + \dots,$$
where m is a positive integer.

102. Prove that if $x^{(-m)}$ have the usual meaning assigned to it, then $\Delta x^{(-m)} = -mx^{(-m+1)}$; and deduce $\Delta^m x^{(-m)}$.

Hence, or otherwise, prove that
$$\frac{1}{n} + \frac{x}{n(n+1)} + \frac{x^2}{n(n+1)(n+2)} + \dots$$
$$= e^x \left\{ \frac{1}{n} - \frac{x}{1!(n+1)} + \frac{x^2}{2!(n+2)} - \dots \right\}.$$

103. Find the sum to infinity of the series
$$1 + 1/(m+1) + 1.2/(m+1)(m+2)$$
$$+ 1.2.3/(m+1)(m+2)(m+3) + \dots.$$

104. Find the value of

$$a^{n+1} - n(a+b)^{n+1} + n_2(a+2b)^{n+1} - n_3(a+3b)^{n+1} + \ldots,$$

where $n_r = n(n-1)(n-2)\ldots(n-r+1)/r!$.

105. Prove that $\Delta^{-1}n_{r-1} = n_r$ (where n_r has the same meaning as in Qu. 104).

Hence, or otherwise, find the value of

$$1 - 4n + \frac{4\cdot5}{1\cdot2}\frac{n(n-1)}{1\cdot2} - \frac{4\cdot5\cdot6}{1\cdot2\cdot3}\frac{n(n-1)(n-2)}{1\cdot2\cdot3} + \ldots \quad (n > 3).$$

106. If $(1 + x + x^2 + \ldots + x^{2m})^n = a_0 + a_1 x + a_2 x^2 + \ldots,$ prove that

$$a_2 + 3a_3 + 6a_4 + 10a_5 + \ldots = \tfrac{1}{6}\{mn(2m+1)^n(3mn+m-2)\}.$$

107. Prove that $\overset{x=n}{\underset{x=0}{\Sigma}}(a+x)^3$ is the product of $\overset{x=n}{\underset{x=0}{\Sigma}}(a+x)$ and $a^2 + \overset{x=n}{\underset{x=1}{\Sigma}}(a+x).$

108. Show that

$$1.n(n+1) + n.(n-1)n + \frac{n(n+1)}{1.2}.(n-2)(n-1) + \ldots$$

$$= \frac{2(2n+1)!}{(n+2)!(n-1)!}.$$

109. If

$$v_0 = u_0 + u_{\frac{1}{10}} + u_{\frac{2}{10}} + \ldots + u_{\frac{9}{10}},$$

$$v_1 = u_{\frac{1}{10}} + u_{\frac{2}{10}} + u_{\frac{3}{10}} + \ldots + u_{\frac{10}{10}},$$

$$v_2 = u_{\frac{2}{10}} + u_{\frac{3}{10}} + u_{\frac{4}{10}} + \ldots + u_{\frac{11}{10}},$$

prove that, if third differences may be neglected,

$$u_{\frac{x}{10}} = \frac{1}{10}v_0 + \frac{2x-9}{20}\Delta v_0 + \frac{2x^2-20x+33}{40}\Delta^2 v_0.$$

110. If $w_0 = \overset{2t}{\underset{0}{\Sigma}} u_r,\ w_1 = \overset{4t+1}{\underset{2t+1}{\Sigma}} u_r,\ w_2 = \overset{6t+2}{\underset{4t+2}{\Sigma}} u_r,$ and differences of a higher order than the second can be ignored, prove that

$$u_{3t+1} = \frac{1}{2t+1}\left\{w_1 - \frac{t(t+1)}{6(2t+1)^2}\Delta^2 w_0\right\},$$

and suggest a simple approximation for the coefficient of $\Delta^2 w_0$ in this expression.

111. Deduce $\Delta^{-1}\sin x$ and $\Delta^{-1}\cos x$ from $\Delta\sin x$ and $\Delta\cos x$.

Hence find the value of $\sum\limits_{r=0}^{r=n-1}\cos(\alpha+r\beta)$.

112. Prove that if the interval of differencing be unity then
$$\Delta\tan^{-1}\{(n-1)/n\}=\tan^{-1}\{1/2n^2\}.$$
Sum to infinity
$$\tan^{-1}\left(\frac{1}{2}\cdot\frac{1}{1^2}\right)+\tan^{-1}\left(\frac{1}{2}\cdot\frac{1}{2^2}\right)+\tan^{-1}\left(\frac{1}{2}\cdot\frac{1}{3^2}\right)+\ldots.$$

113. From a certain table of mortality giving the numbers of deaths at quinquennial age-points the number of deaths at age 44 is estimated at 1000, using the figures given for the deaths at ages 32, 37, 42, 47, 52. Without using divided differences, calculate what must have been the tabular number of deaths at age 32, given that the numbers at ages 37, 42, 47, 52 were respectively 869, 952, 1112, 1371.

114. Prove that
$$u_{x:y}=(1+x\Delta_x+y\Delta_y+xy\Delta_x\Delta_y)\,u_{0:0}$$
$$+\{x_2\Delta_x^2+(x+1)_3\,\Delta_x^3+yx_2\Delta_x^2\Delta_y\}\,u_{-1:0}$$
$$+\{y_2\Delta_y^2+(y+1)_3\,\Delta_y^3+xy_2\Delta_y^2\Delta_x\}\,u_{0:-1}$$
as far as third differences.

Find $(1\cdot02125)^7$, given $1\cdot01^6=1\cdot062$, $1\cdot02^2=1\cdot040$, $1\cdot02^6=1\cdot126$, $1\cdot02^{10}=1\cdot219$, $1\cdot03^6=1\cdot194$, $1\cdot03^{10}=1\cdot344$.

115. The following is an extract from a table showing the annual pension to a widow which a single premium of 1 will purchase:

Age of husband next birthday	Age of wife next birthday		
	20	25	30
25	·532	·555	·586
30	·469	·489	
35	·413		

Find the entry in the table corresponding to age of husband 31 next birthday, age of wife 27 next birthday.

116. $a_{30:30} = 13\cdot930$, $a_{30:35} = 13\cdot491$, $a_{30:40} = 12\cdot897$,

$a_{35:40} = 12\cdot581$, $a_{35:35} = 13\cdot111$, $a_{40:40} = 12\cdot125$.

Find $a_{32:37}$ as correctly as the data will allow.

117. $u_{30:20} = 28\cdot92$, $u_{30:25} = 27\cdot22$, $u_{30:30} = 25\cdot35$,

$u_{35:20} = 34\cdot18$, $u_{35:25} = 32\cdot24$, $u_{35:30} = 30\cdot10$,

$u_{40:20} = 40\cdot19$, $u_{40:25} = 38\cdot08$, $u_{40:30} = 35\cdot62$.

Use all the data to obtain $u_{37:28}$.

118. Prove the formulae for two-variable interpolation

(i) $u_{x:y} = u_{0:0} + \frac{1}{2}x\,(u_{1:0} - u_{-1:0}) + \frac{1}{2}y\,(u_{0:1} - u_{0:-1})$, based on five points;

(ii) the six-point formula consisting of the same terms with the addition of

$$\frac{1}{2}x^2\,(u_{1:0} - 2u_{0:0} + u_{-1:0}) + \frac{1}{2}y^2\,(u_{0:1} - 2u_{0:0} + u_{0:-1})$$
$$+ xy\,(u_{1:1} + u_{0:0} - u_{1:0} - u_{0:1}).$$

119. Obtain a Lagrange formula for $u_{x:y}$, given

$$u_{0:0},\quad u_{0:1},\quad u_{1:0},\quad u_{1:1}.$$

Use the formula to find $u_{31:43}$ from the data:

$$u_{30:40} = 14\cdot45,\quad u_{35:40} = 14\cdot05,\quad u_{30:45} = 13\cdot65,\quad u_{35:45} = 13\cdot31.$$

120. Having given the following annual premiums for a joint-life assurance, find $P_{23:32}$ by second difference interpolation in two directions:

$P_{20:30} = \pounds2$	16s.	4d.,	$P_{25:30} = \pounds2$	18s.	10d.,
$P_{30:30} = \pounds3$	2s.	4d.,	$P_{20:35} = \pounds3$	2s.	2d.,
$P_{25:35} = \pounds3$	4s.	2d.,	$P_{30:35} = \pounds3$	7s.	4d.,
$P_{20:40} = \pounds3$	9s.	10d.,	$P_{25:40} = \pounds3$	11s.	6d.,
$P_{30:40} = \pounds3$	14s.	4d.			

121. If $[n]\,u_0$ represent the sum of n terms of the series of which u_0 is the central term, prove that

$$[4]\,[6] \equiv [5]^2 - [1] \quad \text{and} \quad [5] \equiv [3]^2 - [3] - [1].$$

122. Prove Woolhouse's formula:

$$[5]^3\,(-3u_{-1} + 7u_0 - 3u_1) = 125u_0 \text{ approximately}.$$

123. Show that, as far as second differences,

$$\{[3] + [5] - [7]\}\,u_0 = u_0 - 8\Delta^2 u_{-1}.$$

124. If δu_c, $\delta^2 u_c$, ... are central differences of u_c, where $c = \frac{1}{2}(n-1)$, prove that

$$[n]\, u_c = \sum_0^{n-1} u_x = n \left[1 + \frac{n^2 - 1^2}{2^2 . 3!} \delta^2 + \frac{(n^2 - 1^2)(n^2 - 3^2)}{2^4 . 5!} \delta^4 \right] u_c$$

as far as fourth differences.

125. Prove that

$$\Sigma(u_x v_x) = u_x \Sigma v_x - \Delta u_x \Sigma^2 v_{x+1} + \Delta^2 u_x \Sigma^3 v_{x+2} - \dots.$$

126. If $\theta \equiv \dfrac{E^{\alpha+\beta} - E^\alpha}{\beta}$ denote the most general divided difference operator, where

$$\beta \neq 0, \quad \text{and} \quad x^{\overline{n}|} \equiv x(x - n\alpha - \beta)(x - n\alpha - 2\beta)\dots(x - n\alpha - \overline{n-1}\beta),$$

prove that $\qquad\qquad\qquad \theta x^{\overline{n}|} = n x^{\overline{n-1}|}.$ (Steffensen.)

127. If $x^{\overline{n}|}$ have the same meaning as in Qu. 126, find the values of α and β in order that the usual expressions for $x^{(n)}$, $x^{(-n)}$, $x^{[n]}$ and x^n may be obtained.

128. Prove that $\qquad f(x) = \sum_{r=0}^{r=n} \dfrac{x^{\overline{r}|}}{r!} \theta^r f(0).$

129. Prove that if $u_0 = A$, then the solution of the linear difference equation $(E - a)u_x = 0$ is $u_x = Aa^x$, and deduce the solution of the equation

$$u_x = a\Delta u_x.$$

130. If a is not unity, show that

$$u_x = b/(1 - a) + Aa^x$$

satisfies the equation $\qquad u_{x+1} = au_x + b.$

131. Prove that the solution of the equation

$$(E - a)(E - b)(E - c)u_x = 0$$

takes the following forms:

(i) if a, b, c are equal, $u_x = a^x(mx^2 + nx + p)$;

(ii) if b and c are equal, $u_x = ma^x + (nx + p)b^x$;

(iii) if a, b, c are all unequal, $u_x = ma^x + nb^x + pc^x$.

F E

2

DIFFERENTIAL CALCULUS

132. Define the differential coefficient of any function of x with regard to x.

Find from the definition $\dfrac{d}{dx}\,(e^x \log x)$.

133. Find dy/dx in the following cases:

(i) $y = (a^2 - x^2)^{\frac{2}{3}}\,(a^2 + x^2)^{-\frac{1}{3}}$;　　(ii) $y = (2x + 5)\,e^{y/x}$.

134. A point on a curve is defined by means of the variable α thus:

$$x = 3 \cos \alpha - \cos 3\alpha; \quad y = 3 \sin \alpha - \sin 3\alpha.$$

Prove that the angle which the tangent at any point makes with the x-axis is 2α.

135. Find in its simplest form dy/dx, when $(x^m y^n)^{\frac{1}{m+n}} = x + y$.

136. Differentiate

$\tan^{-1}\{2x/(1 - x^2)\}$ with respect to $\sin^{-1}\{2x/(1 + x^2)\}$.

137. Find the second differential coefficient of y with respect to x given that

$$ax^2 + 2bxy + cy^2 = k.$$

138. Find dy/dx, where

(i) $y = (x - \cos x)^{\sec x}$;　　(ii) $y = ae^{-bx} \cos (nx + c)$.

139. Find the value of

(i) $(D^2 - 2D\Delta + \Delta^2) \log x$;　　(ii) $D\Sigma x^2 e^x$.

140. Trace roughly the graph of the curve whose equation is $y^2 = x\,(x^2 - 3x + 2)$, and find the points at which the tangent to the curve is parallel to the co-ordinate axes.

141. $y = \frac{1}{6} \log \dfrac{(x^4 - 1)^2}{x^8 + x^4 + 1} + \dfrac{1}{\sqrt{3}} \tan^{-1} \dfrac{2x^4 + 1}{\sqrt{3}}$.　Find dy/dx.

142. Prove that if $u = e^{\sin^{-1}x}$ and $v = e^{-\cos^{-1}x}$, then du/dv is independent of x.

143. Differentiate

(i) $\log \sin y$ with respect to x, where $y = x^{\log \sin x}$;

(ii) $\log_{10} x$ with respect to x^2.

144. If $x = 2a \sin^2 t \cos 2t$ and $y = 2a \sin^2 t \sin 2t$, prove that $dy/dx = \tan 3t$.

145. $y = \dfrac{m\,(mx + n\sqrt{a^2 - x^2})}{m^2 + n^2}\, e^{\frac{n}{m}\sin^{-1}\frac{x}{a}}$. Find dy/dx.

146. If $e^y + e^{-x} = 2$, prove that
$$d^2y/dx^2 + (dy/dx)^2 + dy/dx = 0.$$

147. Find d^2y/dz^2, where $y = \log_e (1 + x^2)^x$ and $z = \log_a x$.

148. Prove that $dy/dx = (a^2 - x^2)^{\frac{3}{2}}$, given that
$$y = \frac{1}{4}\left[x\,(a^2 - x^2)^{\frac{3}{2}} + \frac{3a^2x}{2}(a^2 - x^2)^{\frac{1}{2}} + \frac{3a^4}{2}\sin^{-1}(x/a)\right].$$

149. $z = u + v$. Find dz/dx in the following cases:

      (i) $u = \sin^{-1} x$; $v = \sin^{-1} u$;

      (ii) $u = a^{x^2}$; $v = x^{a^x}$;

      (iii) $u = \dfrac{\log x}{x}$; $v = \dfrac{du}{dx}$.

150. $x^2 = 2y$. Prove that $D^5 \cos y = (15x - x^5)\sin y + 10x^3 \cos y$.

151. If $z = (g + z)\,v^{2n}$, where $v = 1/(1 + \frac{1}{2}g)$ and n is constant, prove that
$$z\,\frac{dg}{dz} = \frac{g}{1 - nv\,(g + z)}.$$

152. Prove that, if the differential coefficient of a function of t is proportional (for all values of t) to the value of the function, the latter must be of the form Ae^{kt}.

The water contained in a rectangular tank is flowing through an outlet at the bottom of the tank. The rate at which the water escapes (in gallons per second) is proportional at any time to the depth of water in the tank. It is observed that the depth of the water is reduced from 27 inches to 3 inches in 11 minutes.

Find the rate at which the surface of the water was falling (in inches per second) when there were 15 inches of water in the tank. (Given $\log_e 3 = 1\cdot1$.)

153. $x^2 + 2y^2 - 2xy - 1 = 0$. Prove that $D^2y = -1/(2y - x)^3$ and find D^3y.

154. $ds/dx = \sqrt{1 + (dy/dx)^2}$. Prove that, for the curve
$$y = \tfrac{1}{2}\{\log(\sec x + \tan x) - \sin x\},$$
$$ds/dy = 1 + 2\cot^2 x.$$

155. Obtain the values of the first three differential coefficients of y with respect to x, at the origin, for the curve
$$ax + by + cx^2 + dxy + ey^3 = 0.$$

156. $y = 4x^3 - 3x$ and
$$x = \tfrac{1}{2}\left[(z + a\sqrt{-1})^{\frac{1}{3}} + (z - a\sqrt{-1})^{\frac{1}{3}}\right](z^2 + a^2)^{-\frac{1}{6}},$$
where a is a constant. Find d^2y/dz^2 in terms of z.

157. If $a_{\overline{n}|} = \dfrac{1 - v^n}{i}$ and $v = \dfrac{1}{1+i}$, show that
$$d\theta/di = \frac{1 - v\,(1 + ni\theta)\,(1 + ni\theta - ni)}{ni^2},$$
where $\theta = \dfrac{1}{i}\left\{\dfrac{1}{a_{\overline{n}|}} - \dfrac{1}{n}\right\}$ and n is a constant.

158. If $2x = y^2 + 1/y^2$, prove that $(x^2 - 1)\,D^2y + xDy = \tfrac{1}{4}y$.

159. Verify that $y = (1 - x^2)^{\frac{1}{2}}\sin^{-1} x$ satisfies the equation
$$(1 - x^2)\,D^2y - xDy + y + 2x = 0.$$

160. Find d^ny/dx^n, given that $ye^{-ax} + 2nax = n^2 + n + a^2x^2$.

161. If $y = (\sin^{-1} x)^2$, find the limit when $x \to 0$ of $D^{n+2}y/D^ny$.

162. The infinite series
$$u_1 v + u_2 v^2 + u_3 v^3 + \ldots, \quad \frac{u_1}{i} + \frac{\Delta u_1}{i^2} + \frac{\Delta^2 u_1}{i^3} + \ldots$$
are equal.

Prove that $\quad -dv/di = (1 + i)^{-2}$.

Given that $\quad V = v + v^2 + v^3 + \ldots + v^n$

and $\quad\quad\quad vW = -dV/di$,

show that $\quad iW = iV + V - nv^n$.

163. If $y = \dfrac{(1 + x)^{\frac{1}{2}}}{1 - x}$, prove that
$$2(1 - x^2)\frac{d^2y}{dx^2} - (3 + 5x)\frac{dy}{dx} - y = 0.$$

By means of this relationship obtain the first five terms of the expansion of y in ascending powers of x, and state the relationship that exists between the coefficients of x^n, x^{n-1} and x^{n-2}.

164. If $y = \sin n\alpha/\sin \alpha$ and $x = \cos \alpha$, show that
$$(1 - x^2)\, d^2y/dx^2 - 3x\, dy/dx + (n^2 - 1)\, y = 0.$$

165. Obtain the expansion of d^2y/dz^2 in ascending powers of x as far as the term involving x^5, given that $y = \sin^{-1} x$ and $z = \tan^{-1} x$.

166. $x\,(d^2y/dx^2) + dy/dx + xy = 0$. Show that
$$x\,(d^2x/dy^2) - (dx/dy)^2 - xy\,(dx/dy)^3 = 0.$$

167. Prove that, if $x = \sin(y^{\frac{1}{2}})$, then $(1 - x^2)\, D^2y - xDy = 2$.

168. From the relation in Qu. 167, show that, if n is a positive integer,
$$[(1 - x^2)\, D^{n+2} - x\,(2n + 1)\, D^{n+1} - n^2 D^n]\, y = 0.$$

169. If $x^2 = -(1 + y^3)/3y$, prove that
$$\frac{d^2y}{dx^2} = \frac{2\,(y^2 - x^2)}{(y^2 + x^2)^3}.$$

170. If $\delta = \log_e(1 + i)$ and $v = 1/(1 + i)$, show that
$$\frac{d^n}{dv^n}\,(v^{n-1}\,\delta) = -\frac{(n - 1)!}{v}.$$

171. State and prove Maclaurin's theorem for the expansion of $f(x)$ in a series proceeding by ascending powers of x.

Given $y = \dfrac{\log(x + \sqrt{1 + x^2})}{\sqrt{1 + x^2}}$, expand y by Maclaurin's theorem as far as the term involving x^5.

172. By means of Maclaurin's theorem expand
$$x^2/(x^2 - ax - bx + ab)$$
in powers of x, as far as the term in x^3, and deduce the general term.

173. Find the first four terms in the expansion of
$$\log(\sqrt{1 + x} + \sqrt{1 - x})$$
in powers of x.

174. If a_n is the coefficient of x^n in the expansion of
$$(x^2 + 1)^{\frac{1}{2}}\log(x + \sqrt{x^2 + 1})$$
in a series of ascending powers of x, prove that if n is greater than unity
$$(n + 2)\, a_{n+2} + (n - 1)\, a_n = 0.$$

175. If y and z are functions of x, find $\dfrac{dy}{dz}$, $\dfrac{d^2y}{dz^2}$ and $\dfrac{d^3y}{dz^3}$ in terms of differential coefficients with regard to x.

If y is equal to e^x, find the first four terms of the expansion of y in powers of $\dfrac{x}{e^x}$.

176. Expand $\log(1 + \cos x)$ as far as the term involving x^4
   (i) by Maclaurin's theorem;
   (ii) by the use of the series for $\cos x$ in terms of x.

177. Divide a given number n into two parts such that the product of the pth power of one and the qth power of the other shall be as great as possible.

178. Find the maximum and minimum values of
$$\frac{1 + 2x - x^2 + 2\,(x - x^3)^{\frac{1}{2}}}{1 + x^2}.$$

179. $u_{-3} = -43$; $u_0 = 2$; $u_3 = -7$; and u_x has a maximum value when $x = -1$. Find the value for x for which u_x has a minimum value.

180. Find the maximum value of $\dfrac{3}{x-6} + \dfrac{5}{x+6} - \dfrac{16}{2x+3}$.

181. u_x is a rational integral function of x. From the following data find u_{20}: $u_2 = 0$, $u_{11} = 810$, $\sum\limits_{0}^{4} u_x = 10$, u_x has a maximum value when $x = 4/3$.

182. If $x + y = 2c$, prove that $f(x)f(y)$ will be a maximum or minimum when $x = y = c$ according as $f(c)f''(c)$ is less than or greater than $[f'(c)]^2$.

183. A flagstaff 60 feet high has the top one-third part painted white. Find the distance from the foot of the flagstaff at which an observer must stand in order that the white part of the flagstaff may subtend the greatest angle at his eye, the height of his eye above the ground being 5 feet 6 inches.

184. Through a point A on the circumference of a circle of radius r two straight lines are drawn enclosing an angle α. If the

straight lines meet the circle again at B and C, prove that the maximum area of the triangle ABC is

$$r^2 \sin \alpha \, (\mathrm{I} + \cos \alpha).$$

185. Given that $u_0 = -22$; $\Delta u_0 = -\frac{1}{2}$; $u_4 = 138$, and that u_x has a maximum value when $x = -\frac{1}{3}$, find the value of u_2.

186. Prove that the area of the smallest ellipse that can be drawn through the four points whose co-ordinates are $(\alpha, \, \mathrm{o})$, $(-\alpha, \, \mathrm{o})$, $(\beta, \, \gamma)$, $(-\beta, \, -\gamma)$ is independent of β. [The area of the ellipse $ax^2 + 2hxy + by^2 = \mathrm{I}$ is $\pi \, (ab - h^2)^{-\frac{1}{2}}$.]

187. Plot the curve $y \, (a^2 + x^2) = ax^2$, and prove that there are points of inflexion where $x = \pm a \sqrt{3}$.

188. Find the area of the maximum triangle formed by the tangent to the curve $a^2 y = x^3$, the x-axis and the straight line $x = c$.

189. The lower corner of the leaf of a book is folded over so that it just reaches the inner edge of the leaf. If the width of the leaf is w and the length of the outer edge greater than $\frac{3}{4}w$, find the minimum length of the crease.

190. In the smaller segment cut off from the curve whose equation is

$$\frac{x^2}{36} + \frac{y^2}{16} = \mathrm{I}$$

by the straight line $x = \mathrm{I}$ is inscribed the largest possible rectangle the sides of which are parallel to the axes. Show that the area of the rectangle is approximately $18\cdot5$ square units.

191. Over a door of a house there is a window 2 feet high, with the bottom edge of the glass 7 feet above ground level. The door faces a river 34 feet broad whose near edge is 20 feet from a point vertically under the window. Inside the house and 8 feet from the same point a flight of stairs rises with treads 9 inches wide and risers 7 inches high. A is standing on the stairs trying to see through the window as great a width of the river as possible. On which stair should he stand if his eye is 5 feet above the centre of the tread on which he stands?

192. Prove, by the methods of the differential calculus, or otherwise, that

(i) $\underset{x\to 0}{\text{Lt}}\ \dfrac{\sin x - x}{x^3} = -\tfrac{1}{6}$; (ii) $\tfrac{2}{3}\underset{x\to 0}{\text{Lt}}\ \dfrac{xe^x - \log(1+x)}{x^2} = 1$.

193. Show that $\cos \alpha$ is greater than $1 - \tfrac{1}{2}\alpha^2$.

194. Prove that if x is positive $x\,(e^{2x}+1) - e^{2x}+1$ is positive.

195. $f(x) = \dfrac{a^3 \sin^3 x - x^3 \sin^3 a}{x^3 \sin(a-x)}$. Find $\underset{x\to a}{\text{Lt}}\ f(x)$ and $\underset{x\to 0}{\text{Lt}}\ f(x)$.

196. Prove that $\underset{x\to \frac{1}{2}}{\text{Lt}}\ \dfrac{\cos^2 \pi x}{e^{2x} - 2ex} = \dfrac{\pi^2}{2e}$.

197. Evaluate $\underset{x\to 0}{\text{Lt}}\ (1+3x)^{\frac{1-2x}{x}}$.

198. Find the limit as x tends to $\tfrac{1}{2}\pi$ of the expression

$$\frac{(\tfrac{1}{2}\pi - x) \log \sin x}{e^{\cos x} - 1 + \log(1 + x - \tfrac{1}{2}\pi)}.$$

199. $y = \alpha/\sin \alpha$ and $\alpha = \cos^{-1}(1-x)$. Find the limit as α tends to zero of the second differential coefficient of y with respect to x.

200. If u is a function of two independent variables x and y, explain what is meant by

$$\frac{\partial u}{\partial x},\ \frac{\partial u}{\partial y}\ \text{and}\ \frac{\partial^2 u}{\partial x \partial y}.$$

If $x = r \cos \theta$ and $y = r \sin \theta$, find the values of $\dfrac{\partial \theta}{\partial x}$ and $\dfrac{\partial \theta}{\partial y}$ in terms of x and y and verify that $\dfrac{\partial^2 \theta}{\partial x \partial y} = \dfrac{\partial^2 \theta}{\partial y \partial x}$.

201. If $x = r \cos \alpha$ and $y = r \sin \alpha$, and x, y, r, α are functions of t, prove that

(i) $\cos \alpha\ \dfrac{\partial x}{\partial t} + \sin \alpha\ \dfrac{\partial y}{\partial t} = \dfrac{\partial r}{\partial t}$;

(ii) $\cos \alpha\ \dfrac{\partial^2 x}{\partial t^2} + \sin \alpha\ \dfrac{\partial^2 y}{\partial t^2} = \dfrac{\partial^2 r}{\partial t^2} - r\left[\dfrac{\partial \alpha}{\partial t}\right]^2$.

202. $u = \log \dfrac{y(1+\sin x)^{\frac{1}{2}} + y(1-\sin x)^{\frac{1}{2}}}{(1+\sin x)^{\frac{1}{2}} - (1-\sin x)^{\frac{1}{2}}}$. Find the value of

$$2y \frac{\partial u}{\partial y} + \sin 2x \frac{\partial u}{\partial x}.$$

203. $zt^{\frac{1}{2}} = e^{-m}$, where $m = ht + x^2/4kt$, h and k being constants. Prove that $\dfrac{\partial z}{\partial t} + hz = k \dfrac{\partial^2 z}{\partial x^2}$.

204. Given that $(x-a)^2 + (y-b)^2 = c^2$,

$$x = r \cos \theta,$$
$$y = r \sin \theta,$$

where a, b and c are constants, prove that

$$c^2 \left\{ r^2 - r \frac{d^2 r}{d\theta^2} + 2 \left(\frac{dr}{d\theta} \right)^2 \right\}^2 = \left\{ r^2 + \left(\frac{dr}{d\theta} \right)^2 \right\}^3.$$

205. If $\dfrac{x^2}{c+\alpha} + \dfrac{y^2}{\alpha} = 1$, and $\dfrac{x^2}{c-\beta} - \dfrac{y^2}{\beta} = 1$, show that

$$\frac{\partial \alpha}{\partial x} \cdot \frac{\partial \beta}{\partial y} - \frac{\partial \alpha}{\partial y} \cdot \frac{\partial \beta}{\partial x} = \frac{4cxy}{\alpha + \beta}.$$

206. $\mu_x = -d \log l_x/dx$. Obtain the value of μ_{10}, given the table:

Age x	10	15	20	25	30
l_x	100000	98203	96061	93044	89685

207. Prove that if u_x be a function the fourth differences of which are constant

$$12Du_x = u_{x-2} - 8u_{x-1} + 8u_{x+1} - u_{x+2};$$

and hence find an approximate value for the differential coefficient of $\log u_x$ with respect to x, where

$$u_{x-2} = 42\cdot699, \quad u_{x-1} = 40\cdot365, \quad u_x = 37\cdot977,$$
$$u_{x+1} = 35\cdot543, \quad \text{and} \quad u_{x+2} = 33\cdot075.$$

208. Given $u_6 = \cdot778$, $u_7 = \cdot845$, $u_9 = \cdot954$, $u_{12} = 1\cdot079$, find by divided differences the values of u_8 and $\left[\dfrac{du_x}{dx} \right]_{x=8}$.

209. If u_x is a rational integral function of x of the rth degree and divided differences are calculated from the values of u_a, u_b, u_c, u_d, ..., show that the rth divided difference is independent of the values a, b, c, d, ...

If $u_0=23$, $u_3=44$, $u_4=99$ and $u_7=660$, find by divided differences the value of $\dfrac{du_x}{dx}$ when $x=3$.

210. Given the following table of common logarithms, find the value of $d\,(\log_{10}x)/dx$, when $x=500$:

$$\log 490 = 2 \cdot 69020,$$
$$\log 500 = 2 \cdot 69897,$$
$$\log 510 = 2 \cdot 70757,$$
$$\log 520 = 2 \cdot 71600.$$

From your result deduce an approximate value of $\log_{10} e$.

211. If u_x is a rational integral function of the third degree, prove that

$$\left(\frac{du_x}{dx}\right)_{x=0} = \frac{bc+cd+db}{(a-b)\,(a-c)\,(a-d)} \cdot u_a + \frac{ac+cd+da}{(b-a)\,(b-c)\,(b-d)} \cdot u_b$$
$$+ \frac{ab+bd+da}{(c-a)\,(c-b)\,(c-d)} \cdot u_c + \frac{ab+bc+ca}{(d-a)\,(d-b)\,(d-c)} \cdot u_d.$$

Evaluate

$$a^3\,(bc+cd+db)\,(b-c)\,(b-d)\,(c-d)$$
$$-b^3\,(ac+cd+da)\,(a-c)\,(a-d)\,(c-d)$$
$$+c^3\,(ab+bd+da)\,(a-b)\,(a-d)\,(b-d)$$
$$-d^3\,(ab+bc+ca)\,(a-b)\,(a-c)\,(b-c).$$

212. From the following data find $\dfrac{dy}{dx}$ and $\dfrac{d^3y}{dx^3}$ when $x=115$:

x	y
105	17·53
110	20·80
115	24·54
120	28·83
125	33·71

INTEGRAL CALCULUS

213. Given that $u_x = x \log x$, and $v_x = \log (1 + 1/x)$, find whether the following statements are true:

(i) $\Delta u_x = I v_x$;   (ii) $\Delta^2 u_x = I^2 v_x$.

214. y is a rational integral function of x and $D^2 y = x + a/x^2$. When $x = 1$, $y = \frac{4}{3}$, and when $x = a$, $Dy = \frac{1}{2} a^2$. Find the function.

215. Integrate   (i) $x^{\frac{1}{2}} (1 + x^3)$;   (ii) $\dfrac{5x - 1}{3x^2 - 10x + 10}$.

216. Evaluate $\displaystyle\int \frac{a + bx}{\sqrt{(x + c)}}\, dx$ and $\displaystyle\int \frac{x\, dx}{(1 - x)^n}$.

217. A formula that has been proposed for the force of mortality at the young ages is $\mu_x = \dfrac{A}{2M} \cdot \dfrac{10^{-Ax^{\frac{1}{2}} - B}}{\sqrt{x}}$, where $\mu_x = -\dfrac{1}{l_x} \cdot \dfrac{dl_x}{dx}$, A and B are constants, and M is the modulus.

Obtain an expression for l_x.

218. Find $\displaystyle\int (1 + x)^{-4} \log x\, dx$ and $\displaystyle\int x^3 (a + bx^2)^{-\frac{3}{2}}\, dx$.

219. If x is greater than unity, expand $\displaystyle\int \frac{(x^2 + 1)\, x\, dx}{(x^2 - 1)^2 (x^2 + 2)^2}$ in the form

$$a + \frac{b}{x} + \frac{c}{x^2} + \ldots$$

as far as the term involving the fourth power of the variable.

220. $e^{-x} \dfrac{d^2 y}{dx^2} + a \sin x + b \cos x = \frac{1}{2} (a + b)$. Find y in terms of x, given that, when x is 0, y is a and dy/dx is b.

221. Evaluate $\displaystyle\int \frac{x^3 dx}{(x + 1)(x^2 + 1)}$ and $\displaystyle\int (x + a) \log (x + b)\, dx$.

222. Find the value of the integral of $A x^p (B - Cx^2)^{-\frac{1}{2}}$, in the following cases:

(i) $A = 3$, $p = 3$, $B = 1$, $C = -1$;

(ii) $A = 1$, $p = 2$, $B = 4$, $C = 9$;

(iii) $A = -1$, $p = -2$, $B = 1$, $C = 1$.

223. Integrate $\displaystyle\int \frac{1+(1+x)^{\frac{1}{3}}}{1-(1+x)^{\frac{1}{3}}}\,dx.$

224. Evaluate $\displaystyle\int \frac{d\theta}{\sin\theta + \sin p\theta}$ for the values $p=3$ and $p=2$ respectively.

By means of the substitutions indicated, evaluate the following integrals:

225. $\displaystyle\int\sqrt{x^2-a^2}\,dx$, by the substitution $x=\tfrac{1}{2}a\,(e^y+e^{-y})$.

226. $\displaystyle\int \frac{(x^2-a^2)\,dx}{x\,\sqrt{(x^4+3a^2x^2+a^4)}}$, by the substitution $z=a/x+x/a$.

227. $\displaystyle\int \frac{dx}{(x-1)\,\sqrt{(1+2x+4x^2)}}$, by the substitution

$$z-2x=\sqrt{1+2x+4x^2}.$$

228. $\displaystyle\int \frac{dx}{(x-1)\,\sqrt{(1+2x+4x^2)}}$, by the substitution $x-1=1/y$.

229. $\displaystyle\int \frac{x^{\frac{2}{3}}dx}{(2x^2-3)^{\frac{17}{6}}}$, by the substitution $x^2=\dfrac{3}{2-v}$.

230. The formula connecting μ_x, the force of mortality, and $\mathring{e}_x$, the complete expectation of life, is

$$\mu_x=\frac{1}{\mathring{e}_x}+\frac{1}{\mathring{e}_x}\frac{d\mathring{e}_x}{dx}.$$

If $\mu_x=-\dfrac{1}{l_x}\dfrac{dl_x}{dx}$ and $\mathring{e}_x=a\,(c-x)^n$, find an expression for l_x.

231. Establish the formula for integration by parts, and evaluate $\displaystyle\int_0^1 x^3 e^{2x}\,dx.$

232. Prove that $\displaystyle\int_a^b \phi\,(x)\,dx=\frac{a-b}{h-k}\int_h^k \phi\left\{\frac{a-b}{h-k}x+\frac{hb-ka}{h-k}\right\}dx.$

233. Evaluate the definite integrals

$$\int_1^2 \frac{dx}{9x^2-4};\quad \int_3^5 \frac{dx}{\sqrt{(7+6x-x^2)}};\quad \int_1^2 \frac{(2-3x)\,dx}{x^3\,(2+x)}.$$

234. Give the general formula for integration by parts for a definite integral and find the value of $\displaystyle\int_1^a x^m\,(\log x)^2\,dx.$

235. Given that $u_{\frac{1}{2}} = 0.4u_{-1}$, $u_{-\frac{1}{3}} + u_{\frac{1}{3}} = 0$, $\int_{-2}^{2} u_x \, dx = 44$, and

$$\left[\frac{du_x}{dx}\right]_{x=1} + \left[\frac{du_x}{dx}\right]_{x=-1} = 14,$$

find $u_{-.4}$.

236. Evaluate by the usual method $\int_{0}^{2} \frac{dx}{(1-x)^2}$. What do you deduce from the result?

237. Find the value of $\int_{0}^{\log_e 2} \frac{e^x dx}{5e^x - 1}$, given that $\log_{10} 3 = .4771$, $\log_{10} 2 = .3010$, and that the modulus $= .4343$.

238. Given that $u_0 + u_1 + u_2 = 6$, $\Delta u_1 = 17$, $\int_{5}^{8} u_x \, dx = 510$, find $\sum_{5}^{8} u_x$.

239. Prove that

$$\underset{n\to\infty}{\text{Lt}} \left\{\frac{1+n}{3n^2+1} + \frac{2+n}{3n^2+2} + \frac{3+n}{3n^2+3} + \dots + \frac{2n}{3n^2+n}\right\} = \frac{1}{2}.$$

240. Find the limit when n tends to infinity of the series

$$\frac{1}{2n} + \frac{1}{\sqrt{n+1}\,(n+\sqrt{n+1})} + \frac{1}{\sqrt{n+2}\,(n+\sqrt{n+2})} + \dots$$

$$+ \frac{1}{\sqrt{4n-1}\,(n+\sqrt{4n-1})} + \frac{1}{6n}.$$

241. $u_x = (1+a)^x - (1+bx)$, where a and b are positive. If u_x has a minimum value when $x = n$, and if $\int_{0}^{m} u_x \, dx = 0$, prove that

$$2\,(u_{m+n} - u_n) = b^2 m^2.$$

242. The method of integration by parts can be applied to $\int f(x)\, dx$, where $f(x)$ is a single function of x, by treating $f(x)$ as one function, and unity as the other. Apply the method to solve the following problems:

(a) Evaluate $\int (a^2 + x^2)^{\frac{1}{2}} \, dx$.

(b) Prove that $I_p = \frac{2p}{2p+1} I_{p-1}$, where $I_n = \int_{-1}^{1} (1 - t^2)^n \, dt$.

243. Express $\dfrac{1}{(1+x)^3(1-x)^3}$ in partial fractions, and hence, or otherwise, prove that

$$\int_0^{\frac{\pi}{4}} \frac{d\alpha}{\cos^5 \alpha} = \tfrac{7}{8}\sqrt{2} + \tfrac{3}{8}\log(\sqrt{2}+1).$$

244. By a consideration of the integral of $(1-t^4)^{-1}$ between the limits 0 and $1/\sqrt{3}$, find the sum of the series

$$1 + \frac{1}{5 \cdot 3^2} + \frac{1}{9 \cdot 3^4} + \frac{1}{13 \cdot 3^6} + \dots.$$

245. Given that $u_0 = 10$, $u_6 = 145$, $\displaystyle\int_0^2 u_x dx = 31$, $\displaystyle\int_4^6 u_x dx = 195$, complete the series from u_1 to u_5.

246. Find the value of

(i) $\displaystyle\int_0^1 \frac{dx}{x^{\frac{1}{2}} + x^{\frac{1}{3}}}$;   (ii) $\displaystyle\int (1+x^3)^{\frac{1}{2}} \cdot x^{\frac{1}{2}} dx$.

247. u_x is the general term of the series

$$0,\ 4,\ 13,\ 26,\ 41,\ 54,\ 57,\ \dots.$$

Use the approximate relation

$$\int_0^{10} u_x dx = \overset{9}{\underset{0}{\Sigma}} u_x + \tfrac{1}{2}(u_{10} - u_0)$$

to find a value for $\log_e 2$ to two decimal places.

248. Evaluate the definite integrals

(i) $\displaystyle\int_0^2 \frac{dx}{(4+3x^2)^{\frac{3}{2}}}$;   (ii) $\displaystyle\int_0^{\frac{\pi}{2}} \sin^3 x \cos^3 x\, dx$.

249. If $u_n = \displaystyle\int x^n \cos mx\, dx$, show that

$$m^2 u_n = x^{n-1}(mx \sin mx + n \cos mx) - n(n-1)u_{n-2};$$

and prove that

$$\int x^3 \cos x\, dx = (x^3 - 6x)\sin x + (3x^2 - 6)\cos x + k.$$

250. Evaluate $\int_0^{\frac{\pi}{2}} \dfrac{d\theta}{4\cos\theta+3\sin\theta+c}$, according as c has the values 1, 5 or 7.

251. Obtain the indefinite integral $\int \dfrac{dx}{\sin x\,(1+\sin x+\cos x)}$ in the form $\frac{1}{2}\left[\tan\frac{1}{2}x+\log\dfrac{\sin x}{1+\sin x}\right]$.

252. Integrate $xe^{ax}\sin x$ between the limits 0 and $\frac{1}{2}\pi$.

253. Show that $\int_0^1 (1-x^2)^m\,dx = (2^m m!)^2/(2m+1)!$, where m is a positive integer.

254. Find the values of

(i) $\int_0^\infty e^{-x}\sin x\,dx$; (ii) $\int x^4\sqrt{(1+x^{\frac{5}{2}})}\,dx$.

255. By means of the substitution $y=x+1/x$, or otherwise, find the value of

$$\int_1^2 (x^4-1)\,x^{-2}\,(x^4+x^2+1)^{-\frac{1}{2}}\,dx.$$

256. If u_r is a rational integral function of x of the rth degree such that $\dfrac{du_r}{dx}=ru_{r-1}$, find the most general form of u_1, u_2 and u_3, given that u_0 is equal to a.

Prove that $u_0^2 u_3 - 3u_0 u_1 u_2 + 2u_1^3$ is independent of x.

257. If u_x is a function of x which is positive and continuous for all values of x greater than unity and decreases steadily as x increases, show that

$$u_n > \int_n^{n+1} u_x\,dx > u_{n+1},$$

and that the series

$$u_1+u_2+u_3+\dots$$

is or is not convergent according as the value of $\int_1^n u_x\,dx$ does or does not tend to a limit as $n\to\infty$.

Apply this test to establish the convergency or divergency of the series

$$\frac{1}{1^p} + \frac{1}{2^p} + \frac{1}{3^p} + \dots,$$

where p is (i) a positive proper fraction;
(ii) equal to unity;
(iii) greater than unity.

258. Find the value of

(i) $\int \dfrac{x^6 - 1}{x^5 + x^3}\, dx$;   (ii) $\int_0^1 x^2 \sqrt{1 - x}\, dx$.

259. If $u_n = \int_{-a}^{a} (a^2 - x^2)^n \cos bx\, dx$, prove that, if n is a positive integer,

$$b^2 u_{n+2} - 2\,(n+2)\,(2n+3)\,u_{n+1} + 4\,(n+1)\,(n+2)\,a^2 u_n = 0.$$

260. Integrate $\dfrac{x^3 + 4x^2 + x - 1}{(x^2 + 1)\,(x^2 + 2x + 1)}$ between 0 and 1.

261. Show that $\displaystyle\int_0^{\frac{\pi}{2}} f(\sin x)\, dx = \int_0^{\frac{\pi}{2}} f(\cos x)\, dx$, and hence prove that $\displaystyle\int_0^{\frac{\pi}{2}} \cos x\,(\cos x + \sin x)^{-1}\, dx = \tfrac{1}{4}\pi$.

262. If $u_n = \int x^m\,(\log x)^n\, dx$, obtain a reduction formula connecting u_n with u_{n-1}.

Hence, or otherwise, evaluate $\displaystyle\int x^3\,(\log x)^2\, dx$ and $\displaystyle\int e^{4x} x^2\, dx$.

263. Given that the logarithms to base 10 of 2, 3 and e are ·3010, ·4771 and ·4343 respectively, prove that

$$\int_1^3 x \log_e (1 + 1/x)\, dx = 1 \cdot 601 \text{ approximately.}$$

264. Find the value of

(i) $\displaystyle\int_0^{\infty} \dfrac{dx}{(x+1)^2\,(x^2+1)^{\frac{1}{2}}}$;   (ii) $\tfrac{1}{2}\displaystyle\int_0^{\frac{\pi}{8}} e^x \sin 2x\, dx$.

265. Evaluate $\displaystyle\int_0^{2\pi} \int_0^{\infty} r e^{-r^2/c^2}\, dr\, d\theta$.

266. In a certain industry the number of men employed decreases at a rate proportional at any moment to the number employed at that moment, and the output per man increases at a constant rate. Find an expression for the total output in 10 years, if the numbers employed at the beginning and end of the period are N_0 and N_{10} respectively, and the output per man during the first and tenth years is A_1 and A_{10} respectively.

267. Given that u_x is a rational integral function of x, find u_1, knowing that

(i) $[\Delta u_x]_{x=1} - [\Delta u_x]_{x=-1} = 60$;

(ii) $\sum_{-1}^{1} u_x = 39$;

(iii) the area of the curve $y = u_x$ cut off by the curve, the axes and the ordinate $x = 1$ is $9\frac{1}{2}$;

(iv) u_x is a minimum when $x = \frac{1}{2}$.

268. Find the area of the loop of the curve
$$x^2 = \frac{y\,(1 - 2y + y^2)}{3}.$$

269. Rewrite the equation of the curve
$$r\,(\cos\theta + 2\sin\theta)^2 = a\,(\cos\theta + 3\sin\theta)$$
in rectangular co-ordinates.

Hence, or otherwise, find the area enclosed by the curve and the radii from the origin for which $\theta = 0$ and $\theta = \frac{1}{4}\pi$.

270. Prove that the area between the curve $y^2\,(1 - x) = x^3$ and its asymptote is $\frac{3}{4}\pi$.

271. The area of a closed curve $p = f(\psi)$ may be obtained in the form
$$A = \frac{1}{2} \int_0^{2\pi} [p^2 - (dp/d\psi)^2]\,d\psi.$$

Find the area of the curve for which $p = 1 + \cos\psi$.

272. Find the area enclosed between the y-axis and the part of the curve
$$y^2\,(1 + x) = 1 - x$$
which lies to the right of the y-axis.

273. Prove that the area between the curve $xy^2 = a^2 (a-x)$ and its asymptote is $24a^2/5$.

274. Find the area of the loop of the curve $xy^2 = (x-a)^2 (2a-x)$.

275. Prove that the area of the loop of the curve

$$y^2 + x^2 (x^2 - 5x + 6) = 0$$

is $5\pi/8$.

276. The curve $x^2 y + a^2 y - a^3 = 0$ meets the y-axis in P. Find the area of the sector POQ, where O is the origin of co-ordinates and Q is the point whose abscissa is a.

277. Two points A and B are taken on the curve

$$ay = a (ax + a^2)^{\frac{1}{2}} + x^2.$$

The abscissae of A and B are a and $-a$ respectively. Find the area bounded by the curve and the two straight lines joining A and B to the origin of co-ordinates.

278. Prove that Weddle's rule for approximate integration can be obtained by the combination of a 7-ordinate Simpson's formula with a 7-ordinate three-eighths formula, in the ratio $9:-4$.

279. Given that $f(20) = \cdot 508$, $f(30) = \cdot 378$, $f(40) = \cdot 297$, find the area of the curve $y = f(x)$ between the ordinates through $x = 20$ and $x = 40$.

If you were given in addition $f(25) = \cdot 440$ and $f(35) = \cdot 300$, how would your estimate be revised?

280. $u_x = a + bx^2 + xf(x^2)$. Prove that the area of the curve $y = u_x$ between the ordinates through $x = -h$ and $x = h$ is $\frac{1}{3}h(u_{-h} + 4u_0 + u_h)$.

If $u_x = a + bx^2 + cx^4 + xf(x^2)$, obtain a similar formula in u_{-2h}, u_{-h}, u_0, u_h, u_{2h}.

281. Evaluate $6\int_0^{1/\sqrt{3}} (1 + x^2)^{-1} dx$ correct to three decimal places by Weddle's rule.

282. Prove the approximate formula

$$\int_{-5h}^{5h} u_x dx = \frac{5h}{144} [19 (u_{5h} + u_{-5h}) + 75 (u_{3h} + u_{-3h}) + 50 (u_h + u_{-h})].$$

283. Obtain a formula correct to second differences for the sum of the series

$$u_0 + u_{1/m} + u_{2/m} + \dots + u_r,$$

where r is a positive integer, in terms of u_0, u_1, u_2, $\dots$, u_r and the first and second differences of u_0 and u_r.

Transform the formula so obtained in such a way that the differences employed shall not involve the knowledge of values of u_x for values of x greater than r.

284. If y is a rational integral function of the fifth degree in x, prove that

$$\int_{-1}^{1} y\,dx = \tfrac{1}{2}\,(y_1 + y_2 + y_3 + y_4),$$

where the y's are the values of y at $x = -\alpha,\ \alpha,\ -\beta,\ \beta$, and α^2 and β^2 are the roots of the equation

$$45z^2 - 30z + 1 = 0.$$

285. Prove that if third differences are constant

$$\int_{1}^{7} u_x\,dx = \tfrac{3}{25}\,(39u_5 + 12u_0 - u_{-5}).$$

286. Show that $u_{a+h} - u_{a-h} = \tfrac{1}{3}h\,(u'_{a+h} + 4u'_a + u'_{a-h})$, with an error approximately equal to $\tfrac{1}{90}h^5\,[d^5u_x/dx^5]_{x=a}$.

287. Prove the three-eighths rule for approximate integration, and apply the rule to evaluate $\int_{0}^{6} u_x\,dx$, where $u_x = (1 + \cdot 3x)^{-2}$.

Is the approximation a good one? If not, why not?

288. Prove that $\int_{8}^{10} u_x\,dx$ is approximately equal to

$$\tfrac{1}{3}\,[u_8 + 4u_9 + u_{10}],$$

and state to what order of differences the approximate expression is correct.

Apply the formula to find the approximate value of $\int_{8}^{10} \log_{10} x\,dx$ and, by comparing the approximate result with the exact expression for the integral, calculate $\log_{10} e$ to four places of decimals. [Given $\log 2 = \cdot 30103$, $\log 3 = \cdot 47712$.]

289. Use an approximate formula to determine the value of

$$\int_{100}^{130} (x^3 + x^2 + 1)\, dx.$$

290. A tank which holds 100 gallons of water when full is filled every evening at 10 o'clock. The tank has a leak which allows water to escape at a constantly diminishing rate of which the following are observed values:

Time	Number of gallons per hour
At 10 p.m.	10
At midnight	8
At 2 a.m.	6
At 4 a.m.	4
At 6 a.m.	2

Given the following times of sunrise, estimate the total amount of water lost through the leak between 10 p.m. and sunrise during the 15 nights from 10 p.m. on 31 March to sunrise on 15 April.

Date	Time of sunrise
April 1	5.36 a.m.
,, 8	5.21 a.m.
,, 15	5.6 a.m.

291. Prove that $\int_{a}^{a+h} \phi(x)\, dx$ is approximately

$$\tfrac{1}{4}h\, [\phi(x_1) + \phi(x_2) + \phi(x_3) + \phi(x_4)],$$

where

$$x_1 = a + h/10, \quad x_2 = a + 4h/10, \quad x_3 = a + 6h/10, \quad x_4 = a + 9h/10.$$

292. The following data are given:

x	1·0	1·1	1·2	1·3	1·4	1·5	1·6
y	1·218	1·504	1·820	2·169	2·554	2·978	3·444

Find the value of

$$\int_{1}^{1\cdot4} y\, dx - (dy/dx)_{x=1\cdot4} + (dy/dx)_{x=1}.$$

PROBABILITY AND MEAN VALUE

293. Three cards are drawn from an ordinary pack of 52 cards. Find expressions for the respective probabilities of the following results on the assumption that the cards are drawn singly and replaced:

(1) A spade each time;
(2) No spades at all;
(3) One spade and two other cards;
(4) At least one spade;
(5) Three cards of the same suit;
(6) Three cards of different suits;
(7) Three aces;
(8) No aces;
(9) All red cards;
(10) Two cards of the same colour and the other different.

294. A put 10 white and 10 black balls in a bag, handed the bag to B without letting him know how many balls there were of each colour, and asked him to make 1000 drawings, replacing each ball drawn after noting its colour. Without the knowledge of A or B, C examined the contents and substituted another black ball for one of the white balls before the drawings commenced. The result of the 1000 drawings was White 470; Black 530.

A then says that the probability of drawing a white ball from the bag is $\frac{1}{2}$, and that the actual result was an accidental error. B says that the probability is $\frac{47}{100}$. C says that it is $\frac{9}{20}$, and that the result was an accidental error.

Who is right and why?

295. Seventeen persons are to be seated at a round table. Three of these persons A, B, C are chosen at random. Find the chance that no two of the three are seated next to each other.

296. A shallow box is in the form of an equilateral triangle of side 4 inches. Sixteen pieces of wood, each an equilateral triangle of side 1 inch, fit into the box. Two of these pieces are red and the rest are black.

(i) How many different patterns can be made?

(ii) If the arrangement is at random, what is the chance that the two red pieces touch each other along a side?

297. Of three independent events the chance that the first only should happen is $\frac{1}{4}$; the chance that the second only should happen is $\frac{1}{8}$; and the chance that the third only should happen is

$\frac{1}{12}$. Show that the independent chance of the second event is $\frac{1}{3}$, and find the independent chances of the other two events.

298. Of all the employees of a certain company, one in three is a woman and one in twelve is a married woman. One in six of the men is a married man. Fifty per cent. of the single women and half this percentage of the married women are members of the company's provident scheme. The percentage of the married men who are members of the scheme is twice the percentage of the single men who are members. The ratio of the total number of employees who are members of the scheme to those who are not is as 7 to 9. Six married men and 5 per cent. of the married women are married couples and they are all members of the scheme.

What is the probability that a person picked at random from among the employees is a married male member of the scheme whose wife is not an employee?

299. A pack of 52 cards is dealt in the usual way to four players. One player has just five cards of a particular suit. Find the chance that his partner has the remaining eight cards of that suit.

300. Four coins are tossed together, and A is to receive £2 if exactly two heads turn up and to pay £1 in any other event. Find the probability that after four trials A is £1 out of pocket.

301. Four different articles are distributed to n persons with no restrictions as to how many any person may receive. Find the probability that two will be given to one person and one to each of two others, for the cases when (i) $n = 4$, (ii) $n = 5$.

302. Each of three bags contains 100 balls, some red, some white, and some blue. If one ball is drawn from each bag, the following are the respective chances of certain results:

Result				Chance
A red ball from bag No. 1	...	...	...	6/10
A red ball from bag No. 3	...	...	...	5/10
A white ball from bag No. 2	...	...	...	4/10
A white ball from bag No. 2 only		...	...	224/1000
A blue ball from each of the three bags			...	6/1000
A blue ball from each of bags Nos. 1 and 2 only			14/1000

Find the number of balls of each colour in each bag.

303. Three boxes each contain nine counters, numbered respectively from 1 to 9. A man draws a counter at random from each box. Find the chance that the sum of the numbers on the counters thus drawn is 24.

How would his chance be altered, if at all, if the 27 counters were all put in one box and the man drew three counters at random?

304. If a coin be tossed 1000 times, how many times would you expect to get a sequence of exactly five heads?

305. What is the probability that an even whole number consisting of four digits is a perfect square?

306. There are five bags containing coloured balls, as follows:

Number of balls of specified colour

Bag No.	White	Black	Red	Green	Yellow
1	2	1	2	1	2
2	2	4	2	0	0
3	3	3	2	0	0
4	1	2	3	1	1
5	3	1	4	0	0

A draws one ball from bag No. 1, B one from bag No. 2, C one from bag No. 3, D one from bag No. 4 and E one from bag No. 5. Calculate the probability that the balls are all of different colours.

307. If a coin is tossed 12 times, what is the probability of getting heads exactly twice as many times in the first eight throws as in the last four?

308. There are $2m$ black balls and m white balls from which six balls are drawn at random. Prove that, if m is very large, the chance of drawing four black and two white balls is $\frac{80}{243}$; and the chance of drawing two black and four white is $\frac{20}{243}$.

309. If a single throw of three dice be made, what is the probability that the total number shown is not less than five times the lowest number shown on any one of the three dice?

310. Five dice are thrown three times in succession. Calculate the probability that in at least one of the throws there will be three or more of the five dice showing the same number.

311. Distinguish between the absolute probability of an event, the average number of times that it may happen over a series of trials, the most probable number of times that it will occur, and the most probable value. Illustrate your answer by reference to probabilities associated with a single throw of two dice.

312. A prize is to be won by A as soon as he throws 5 with two dice, or by B as soon as he throws 10 with three dice. They throw alternately, A commencing. Find their respective chances of winning the prize.

313. A is one of 32 players who enter for a tournament. The players are drawn in pairs and the winners of each round are drawn again for the next round. All the players are of equal skill. Find the chance that A will either win or will be beaten by the actual winner.

314. A man throws 10 dice and removes all that show a 6; he then throws the remainder and removes all that show a 6; and so on. How many times ought he be allowed to repeat this operation if he is to have an even chance of removing all the dice before he has finished?

315. A and B play a game as follows. From an ordinary pack of 52 cards the court cards (J, Q, K) are removed. The remaining cards are separated into suits, and the four packs thus formed are placed face downwards on the table. A turns up the top card of each pack and notes the total of his cards. He replaces the cards, each pack is shuffled separately and B turns up the top card of each pack. B notes his total. The process is repeated until A's total is exactly 22 or B's total is exactly 23, whichever happens first.
Find A's chance of winning the game.

316. A and B toss a penny in turn, A beginning. Heads count 3, tails 1. The tossing continues until A tosses a head, B then having one more toss to complete the game. A wins if his total is not less than B's. What odds should A give B to make the game fair?

317. A and B play a match which is won by the player who first wins two games. The chances that A wins, draws or loses any game are a, b and c respectively. Find A's chance of winning the match.

318. There are 10 counters in a bag, each marked with one of the numbers 0 to 9. A man draws a counter and notes the number on the counter. If he draws 3 or 7, he replaces the counter and no score is credited to him. If, however, he draws a counter with any number except 3 or 7, the number is added to his score and the counter is not replaced. He draws three times in succession, a draw which produces a 3 or 7 being reckoned as one of his turns. Find, correct to three places of decimals, his chance of scoring exactly 3.

319. If two dice be tossed three times, what is the probability that the totals shown on the three occasions taken in order of throw will be in arithmetical progression with a common difference greater than two?

320. If there are three bags, one containing three white balls and three black balls, another containing three white balls and three red balls, and the third containing three red balls and three black balls, and two balls are drawn from each bag, what is the probability that there will be at least one ball of each colour among those drawn?

321. A certain rifleman's chance of scoring a bull is $\frac{1}{2}$ if he scored a bull with his previous shot or $\frac{1}{3}$ of his chance at the previous shot if that was not a bull. In each practice he ceases fire when he has had five shots or has scored two consecutive bulls, whichever occurs first. If his chance of scoring a bull at his first shot in each practice is $\frac{1}{4}$, find correct to three places of decimals his average number of shots per practice.

322. There are 30 counters in a box, exactly alike except as regards colour; 10 are white, 10 black, and the other 10 are of different colours (not black or white). Find the chance that if 10 counters are drawn together at random none of them will be black or white.

323. Three black balls and five white balls are placed in a bag. Three men draw in succession until a black ball is drawn. Find their respective chances,

(i) if the drawn ball is continually replaced before the next drawing;

(ii) if the drawn ball is not replaced.

324. A and B play a series of games, the winner of each game scoring one point. If a game is drawn, each player scores $\frac{1}{2}$. In each game the probabilities that A wins, draws, or loses, are $\frac{1}{2}$, $\frac{1}{4}$, and $\frac{1}{4}$ respectively. Find the chance that after five games have been played neither player will have scored more than three points.

325. Thirteen players enter for a competition on the knock-out principle. The opponents for the first round are drawn in pairs by lot, the odd man out being a bye. The six winners of the first round and the bye are then drawn again in pairs, the odd man out being a bye. The three winners of the second round and the bye are then drawn in pairs for the semi-final round. On the assumption that all the players are of equal skill, find the probability that two brothers who enter for the competition are opposed in the final round.

326. A and B play a game with an ordinary pack of 52 cards against Y and Z. Each of the four players has 13 cards, and B's cards are exposed face upwards.

(i) If A has five hearts and B four, find an expression for the chance that the remaining hearts lie three in one hand and one in the other.

(ii) If A has six hearts and B three, find an expression for the chance that the remaining hearts lie two in one hand and two in the other.

327. If three dice be thrown, what is the probability that the difference between the highest and lowest numbers shown on the three dice is more than 3?

328. In a certain community the proportion of persons suffering from colour-blindness is and has been at all material times 1 in 100. The average number of children in a family has been four for the last two generations. What is the probability that if a member of the community be chosen at random, at least one of his grand-parents, parents, aunts, uncles, cousins, brothers and sisters is colour-blind? It may be assumed that no marriage takes place between the remaining members of his mother's and his father's families.

Give your result correct to two places of decimals without the use of tables.

329. A penny is to be tossed seven times. What is the probability that there will be at least one sequence of three or more heads?

330. Cards are turned up one by one from an ordinary pack of 52 cards until an ace appears. Find the chance that an ace does not appear in the first 26 cards.

331. A bag contains 8 black balls and 2 white balls. One ball is drawn at a time, and any that are black are replaced in the bag and any that are white are set aside. What is the average number of drawings required to draw both the white balls?

332. A particular Army Corps consists of 24 Battalions, each consisting of 1000 men. It is to be divided into 4 Divisions each to consist of 6 Battalions, and it is equally likely that any particular Battalion will be allotted to any one of the Divisions. What is the probability that 100 men, known to be serving in the Army Corps, who were only eligible to enlist in 2 particular Battalions out of the 24, and each of whom is equally likely to have enlisted in either of the 2, will all be in the same Division?

How much, if any, of the information given above has no bearing on the probability in question?

333. A, B and C play a series of three matches. The scoring is as follows. Six points are given for each game, the winner scoring four points and the runner-up two points. If two players are equal for the first place, they each score three points; if two players are equal for the second place, they each score 1 point, the winner scoring four points as before. For a tie between the three each scores two points. Assuming that any one of the possible results of a match is as likely as any other, find the chance that at the end of the series of three matches any one player has scored six points.

334. Given that x and y are positive whole numbers whose sum is 120, find the probability that the product xy will exceed half its maximum value.

335. Out of n occasions in which an event of probability p is in question, on what number of occasions is it most likely to happen?

Apply the result to find the most likely number of aces to appear when an ordinary die is thrown 20 times.

336. A man is to gain a shilling on the following conditions. He draws twice, replacing each time, out of an urn containing one white and one black ball. If he draws the white ball twice, he wins. If he fails to draw a white ball, a black ball is added; he tries twice again and wins if he draws the white ball twice. If he fails, another black ball is added, and so on. What is his chance of gaining the shilling?

337. Distinguish between "probable value" and "most probable value".

A man tosses a coin 20 times. For every sequence of m heads or m tails he is to receive $2^m - 1$ shillings, m being not less than 10. Find his expectation to the nearest shilling.

338. In a certain competition four of the six letters A, B, C, D, E, F have to be selected and arranged in order. The prize is awarded to the person whose arrangement of four letters agrees with an order obtained by drawing four letters one by one at random from the six letters. If more than one person is successful, the prize is shared equally. If no one has the correct arrangement, those with only one letter wrong then rank equally for the prize.

If there are 1000 competitors, what is the probability that a certain competitor will win the whole prize with an arrangement that has one letter wrong?

339. A and B each throw two dice. If A throws 7 before B throws 4, B pays A two shillings; if B throws 4 before A throws 7, A pays B five shillings. They throw simultaneously until one of them wins, equal throws (i.e. A 7 and B 4) being disregarded. Find B's expectation.

340. A and B are tossing a coin in turn. Heads count 3, tails 1, and the first to score 10 or more wins the game. The score is A 7, B 8, A to play, B having tossed last and scored 3. A questions whether B's last toss was really a head, and asks B to cancel his 3 and toss again. B proposes instead that he should score 2 for his last toss, pointing out that that is the expectation at each toss. Prove that if A agrees, his chance of winning will be less than it would be if his proposal were adopted.

341. In a dice game in which three dice are thrown, the thrower selects any number from 1 to 6 and receives an amount equal to his original stake if the number selected turns up on one die, twice his original stake if the number turns up on two dice, or three times his stake if the number turns up on the three dice. In any other event he loses his stake. If the dice are thrown twice each minute, and a man starts with £5, staking a sum of a shilling on one number at each throw, how long may he expect to play before his fund is reduced to £2. 10s. 0d.?

342. A domino is a rectangular piece of wood, one face of which bears two of the numbers 0, 1, 2, 3, 4, 5, 6 repeated if necessary. For example a domino may bear 0, 0; 3, 4; 5, 5; ... (but 4, 3 is not different from 3, 4).

Two dominoes are drawn at random from a set of dominoes marked with all the possible variations of the numbers above. Find the chance that the sum of the numbers on the two dominoes thus drawn is a multiple of 5.

343. The probability that at least one of three persons all aged x will die within n years is $\frac{19}{27}$. Find the chances that, of four persons all aged x,

(i) At least two will attain age $x+n$;

(ii) One and only one will attain age $x+n$;

(iii) One at least will die within n years.

344. In the case of three lives aged 20, 55 and 65, the probability that exactly one survives 10 years is ·121875, and that at least one dies within 10 years is ·578125. Find the chances that a life aged 65, a life aged 55 and a life aged 20 each survives 10 years, given that these chances are in geometrical progression.

345. A man throws an ordinary six-faced die five times, and scores the reciprocal of the number that turns up. What is the chance that his total score will be a whole number?

346. A, B, C each throw two dice for a prize. A is to throw 3 or 11 to win, B is to throw 7 to win and C is to throw any number between 2 and 12 inclusive (except 3, 7 or 11) to win. Directly the winning number is thrown the game ceases. A starts and is allowed

three consecutive throws; if he fails to throw 3 or 11, B throws. He is allowed two consecutive throws, and if he fails to throw 7, C throws. C has one throw only, and if he fails to throw one of the numbers allotted to him A throws again, the succession being A three throws, B two throws, C one and so on. Which of the three has the best chance of winning? If the prize is 30s., find A's expectation to the nearest shilling.

347. A bag contains 10 counters—three red, three black, two green, two white; a second bag contains 11 counters—three red, four black and four white. Each counter has a value—red 1s., black 6d., green 3d. and white nil. A man draws a counter at random from the first bag and places it in the second; he then draws one at random from the second bag and places it in the first. Assuming that any one counter is as likely to be drawn as any other, find, after the second draw,

(i) the chance that one green counter only is in the first bag;

(ii) the most probable value of the first bag;

(iii) the probable value of the first bag.

348. A bag contains five counters marked 1, 2, 3, 4, 5 respectively. A counter is drawn and replaced; this is repeated a second and a third time. A man is to receive a number of shillings equal to the sum of the numbers drawn and, in addition, a number of pounds equal to the sum of the numbers drawn when they are three consecutive numbers in their natural order. What is the value of his expectation?

349. Suppose that of 100 persons born together one dies annually until all are extinct: find the probability that of four persons aged respectively 20, 30, 55 and 70, three at least will be living 15 years hence.

350. A group of r houses is occupied by families whose members may range in number from 1 to r. Families numbering 1, 2, 3, 4, ..., r are equally likely. What is the chance that the average number of members per house is exactly three?

351. There are three purses A, B, C, each of which contains counters exactly alike except as regards colour. White counters are

worth 1d., red 2d., blue 3d. Initially the contents of the purses are A, five white, three red, two blue; B, two white, three red, one blue; C two white, five red, one blue. Two counters are taken at random from A and placed into B; two are then taken at random from B and placed into C; two are then taken at random from C and placed into A. This process is repeated. Find the probable value of purse C at the end of the second series of replacements.

352. Given the following probabilities, find the chance that out of four persons aged exactly 96, the third to die will die at age 97 last birthday.

Exact Age	Chance of dying before next birthday
96	$\frac{1}{2}$
97	$\frac{2}{3}$
98	$\frac{3}{4}$
99	1

353. A is approaching on foot a street-crossing 30 feet wide at 100 yards per minute, followed by B who is walking at 120 yards per minute. When A is 100 yards from the crossing, B is 100 yards behind A. At the crossing, automatic traffic signals allow foot-passengers 30 seconds clear out of every 3 minutes in which to begin to cross the road, a few seconds warning being given that the 30 seconds period is about to begin or has ended. Assuming that neither A nor B will begin to cross the road except during the 30 second clear period, what is the probability that B will overtake A before he reaches the other side of the road?

354. A, B, C, D are each exactly 34 years old. The chance that exactly two of them survive to age 51 is $\frac{8}{27}$. Find the chances that,

(i) D will die before age 51 and will be survived by only one of the others;

(ii) they all die before age 51, their deaths occurring in the following order:

(1) either A or B;

(2) C;

(3) the survivor of A and B;

(4) D.

355. Six football teams including X, Y and Z enter a tournament. The teams are of equal strength and are drawn by lot in pairs before the next round, the winners of the previous round entering the next round. If there is an odd number of contestants in any round, the team drawing a bye also enters the next round. Find the chance that in the course of the competition X will defeat Y after having first defeated Z.

356. There are three sections to an examination paper. Section A contains three questions; Section B four questions; Section C five questions. The maximum marks awarded to the questions are 30, 30, 40 for the three in A; 30, 30, 40, 45 for the four in B; 30, 30, 40, 40, 45 for the five in C. A candidate is as likely to choose any one question as any other, and the maximum marks for each question are not disclosed to him. If the instructions are that one question must be chosen from A, two from B and two from C, find the chance that a candidate who chooses five questions chooses them so that the maximum attainable is at least 180.

357. The chance that two men, one aged 20 and the other aged 40, both live 20 years is ·61177. Of 96,223 men alive at age 20, 6358 die before they attain age 30. Find the chance that a man aged 30 will live to reach age 60.

358. A has to appear in Court as a witness, and unless he can leave by 11.30 a.m. he will be late for another appointment. There are four other witnesses, the first of whom will be called at 10.30 a.m., and the time to be occupied by each witness is as follows:

A 15 minutes, B 30 minutes, C 10 minutes, D 10 minutes, E 5 minutes.

Except that A will not be called before B, the witnesses may be called in any order, all orders subject to this restriction being equally likely. What is A's chance of leaving by 11.30 a.m.?

359. A pack of cards is dealt in the usual way to four players so that each player has 13 cards. It is known that two of the players hold six hearts between them. Find the chance that each holds three hearts.

360. From a bag containing p white balls and q black balls, one ball is drawn and replaced, this operation being performed n times.

If Q_n is the probability that no two consecutive draws produce two white balls, find the relationship between Q_n, Q_{n-1} and Q_{n-2}.

Find from this relationship the value of Q_n when $p=4$ and $q=9$.

361. A point is scored by X each time that he throws a 6 with one die, and a point is scored by Y each time that he throws 6 with two dice. X engages to score 10 points more than Y before Y scores 2 points more than X. Find X's chance of winning the match.

362. An experiment succeeds three times out of four. If u_n be the chance that in n consecutive trials there are never three consecutive successes, show that

$$u_0 + u_1 x + u_2 x^2 + \ldots + u_r x^r + \ldots$$

is a recurring series, and find the scale of relation.

If $n=6$, find the value of u_n.

363. A private code is to consist of a number of three digits, repetitions being allowed. The following conditions are imposed:

 (i) A zero may occur anywhere except at the beginning;

 (ii) Numbers containing two or more consecutive digits in their natural order are inadmissible, the natural order being 0, 1, 2, ..., 9.

A code-number is chosen at random. Find the probability that it contains at least one zero.

364. Twelve persons, three of whom are brothers, are travelling by train. They find that there are only three compartments available for them. Assuming that they enter the three compartments at random, except that not less than three enter each compartment, calculate the probability that the three brothers are all in one compartment.

365. A bag contains 20 counters; they are exactly similar except that 19 are numbered consecutively from 1 to 19, while the other one is marked with the number 21. A man draws two counters at random. If the sum of the numbers on the counters is odd he is to receive that number of shillings, but if the sum is even he is to pay that number of shillings. Show that his expectation is about $9\frac{1}{2}d$.

366. If four numbers be chosen at random from the first $2n$ natural numbers, find the chance that the sum of two of the numbers thus chosen is equal to the sum of the other two.

367. Penny tickets can be obtained from an automatic machine by inserting one penny or two halfpennies. On clearing the machine it is found that it contains eight pennies and eight halfpennies. What is the probability that at no time since it was previously cleared did it contain more halfpennies than pennies?

368. Three players play the following game. Ten cards, each bearing one of the numbers 1 to 10, are shuffled and three cards are dealt to each player. The tenth card is put aside face downwards. Each player stakes 1s., and it is agreed that the player whose cards add up to the highest total shall take the 3s., the money to be divided if two or more players have equal totals. One of the players has the 9, the 6 and the 2. Find his expectation.

369. A, B and C play at a game in which each has a separate score, and the game is won by the player who first scores two points. If the chances are respectively $\frac{1}{2}$, $\frac{1}{3}$, $\frac{1}{6}$ that any given point is scored by A, B, C, find the respective chances of the three players winning the game.

370. Omnibuses going along three different routes stop at a point A. A man is at A at 12 noon, and wishes to be at a point B not later than 12.30. If he takes bus No. 1 he can travel to a point C 10 minutes journey away, and from C he can walk to B in 16 minutes. By bus No. 2 he can travel to D in 15 minutes; from D he can walk to B in 8 minutes. He can ride all the way by bus No. 3 in 20 minutes. Assuming that he takes the first bus that comes along—which in any case will not be later than 12.10—find the chance that he arrives at B not later than 12.30. All times between noon and 12.10 and any bus are equally likely.

371. It has been stated that 926 people out of every 1000 catch at least one cold every year. If this is correct, how often on the average does a person catch a cold, assuming that the chance of catching a cold at any instant is constant throughout the year?

372. A bag contains n balls, three red and the rest white. They are drawn out one by one. Find the probability that no two red balls will be drawn consecutively.

373. There are $2m$ black balls and m white balls. Six are drawn at random. Find the chance that, when m is very large, four of the balls are black.

374. A penny is tossed six times. Prove that the chance that neither heads nor tails have occurred three times in succession is $\frac{13}{32}$.

375. The correct weight of a certain coin when newly minted is 123·274 grains and the permissible error in the weight is plus or minus ·2 grain. Calculate the probability that the weight of two newly minted coins taken at random will be less than 246·500 grains, assuming that all errors between the permissible limits are equally likely.

376. An eight-page pamphlet is being printed in one operation on a single sheet of paper, four pages on each side, the paper being subsequently folded, stitched and cut. Each page is printed from a separate cast. If the casts are put on the printing machine so as to print the correct way up, but otherwise in any order at random, what is the probability that the paper can be folded so that the pages of the finished pamphlet will be in numerical order?

377. Three curves A, B, C are drawn on the same axes OX, OY. Their equations are

$$y^2 = 4ax \ (A); \quad y^2 + 4ax = 8a^2 \ (B); \quad y^2 = 4a \ (x-a) \ (C).$$

A and B intersect at L and M; B and C intersect at H and K. If B and C cut the x-axis at P and Q respectively, find the chance that a point taken at random in the figure $OLPM$ falls within the figure $QHPK$.

378. AB is a straight line 12 inches long. Two lengths, one of 8 inches and the other of 4 inches, are cut off at random. Find the chance that the common part is not greater than one-quarter of the line.

379. A ship steering a straight course at a uniform speed picked up a stationary object 2 miles away with the beam of a searchlight, the beam making an angle of 45° ahead with the course of the ship. The beam was kept on the object until it made an angle of 60° astern with the course. Calculate the mean distance of the object during this time.

380. If a point be taken at random within a square, find the probability that none of the straight lines joining it to the four corners of the square will be longer than the side of the square.

381. Within the hour between 2 a.m. and 3 a.m. a light signal is to be shown at a certain place for 1 minute. If A passes by so as to have the place in view for five successive minutes during this hour, what is the probability that he will observe the signal for some part of its duration?

382. A sets his watch every evening by wireless time-signal at 9 p.m. exactly. He finds that his watch gains 1–3 minutes per day, all times between these limits inclusive being equally likely. One day, at 1 minute past 12 noon by his watch, he leaves his office.
What is the probability that he leaves before 12 o'clock?

It may be assumed that the time gained by the watch in any one day is uniformly distributed over the 24 hours, and changes from ordinary time to summer time and *vice versa* may be ignored.

383. Two lengths, PQ and MN, each 6 inches long, are measured at random on a straight line 16 inches long. Find the chance that no point on PQ coincides with a point on MN.

384. A straight line is divided into two parts at random, and one part is then divided at random into two parts. Find the chance that the three parts of the line can be put together so as to form a triangle.

385. Three events A, B, C are known to have happened in the same century. What is the chance that the events happened in the order A, B, C; B happening within n years of the middle of the century?

386. An explorer in the desert is looking for a well which he will be able to identify at a distance of 3 miles. At a moment when he

is actually 5 miles from the well he decides to walk in a circle of 3 miles radius. If he may start off in any direction, what is the probability that he will come within sight of the well before he completes the circle?

387. A point is selected at random on the circumference of a circle of radius r. Find the mean value of the distance between this point and a fixed point on the circumference of the circle.

388. Show that

$$\sin x = x - \frac{x^3}{3!} + \frac{x^5}{5!} - \dots$$

A large table is divided by parallel lines 1 inch apart into strips coloured alternately black and red. If a coin 1 inch in diameter is dropped at random on to the table, find to three places of decimals the chance that the part of the coin lying over a red strip is less than $\frac{1}{2}$ per cent. of the total area of the coin. [$\sqrt[3]{\cdot 06\pi} = \cdot 573$.]

389. A point is taken at random in the rectangle formed by joining the points $(2c, \sqrt{3}c), (2c, -\sqrt{3}c), (-2c, \sqrt{3}c), (-2c, -\sqrt{3}c)$. Find the chance that it falls within the space bounded by the curve $x^2 - y^2 = c^2$ and the sides of the rectangle parallel to the x-axis.

390. A table giving for each integral age the force of mortality according to the formula $\mu_x = A + Bc^x$ is entered inversely with a value μ lying between μ_y and μ_{y+1}, and the age to which the value corresponds is determined approximately by the ordinary method of proportional parts to be $y + \alpha$. If the true age be $y + \beta$, find

(i) the maximum value of $\beta - \alpha$;

(ii) the mean value of $\beta - \alpha$, if all values of β between 0 and 1 are equally likely.

391. Through the mid-point of a radius (r) of a circle a straight line is drawn at right angles to the radius, thus dividing the circle into two unequal parts. From all points on the base of the smaller segment straight lines are drawn at right angles to the base, terminating on the circumference. Find the mean length of these lines.

392. From a point A on the circumference of a circle, two straight lines are drawn to two other points B and C taken at random on the circumference. Find the mean difference in length between the lines AB and AC.

393. AB and CD are perpendicular diameters of a circle. Find the mean value of the distance of A from points on the semicircle CBD, and the mean value of the reciprocal of that distance. (Given $\tan \pi/8 = \sqrt{2} - 1$.)

394. Calculate the mean value of the ordinates of a semicircle according as the points from which the ordinates are drawn are taken at equal distances

   (i) along the diameter;

   (ii) along the circumference.

395. Find the mean value of (i) the ordinates, and (ii) the squares of the ordinates from points taken at equal distances along the x-axis from $x = -a$ to $x = a$, drawn to the curve $x^2/a^2 + y^2/b^2 = 1$.

396. $ABCD$ is a square of side a. Along AB any point H is taken, and along BC, CD, DA points K, L, M are taken such that BK, CL, DM and AH are equal. AK, BL, CM, DH are joined intersecting one another to form an inner square $PQRS$. Find the mean value of the area of the square $PQRS$.

397. A says goodbye to B at X with the intention of walking by road to Z. From X the road runs for 1 mile in a straight line to Y and then turns at right angles running in a straight line to Z. Four minutes after A has started B sets off from X in a straight line across country to the road YZ with the intention of waylaying A.

If B proceeds at 3 miles an hour, and A may proceed at any uniform rate from 3 to 4 miles an hour, all rates between these limits being equally likely, to what point on the road YZ should B proceed in order to have the best chance of catching A and what will this chance then be?

398. P, Q, R, S are four villages in a straight line, the distances PQ and RS being each half the distance QR. A man sets out to walk from P through Q and R to S at some time between noon and

1 p.m.; a cyclist starts from S in the reverse direction at some time between 12.40 p.m. and 1 p.m. If the speed of the cyclist is three times that of the man, and if the man would normally take an hour to cover the whole journey, find the chance that they meet between Q and R. All times of starting between the given limits are equally likely, and the man and the cyclist are assumed to travel throughout at uniform speeds.

399. A, B and C run a mile race. A runs at $20\frac{1}{2}$ feet per second, B at 20 feet per second, and C at $19\frac{1}{2}$ feet per second. A starts from scratch, B has 50 yards start, and C 80 yards start. Find the mean distance between the leading man and the last man during the time that elapses from the start of the race until the winner reaches the winning post.

400. $ABCD$ is a rectangle. $AB = 2b$; $BC = 3b$. P is the midpoint of AB, and Q is a point on BC such that $BQ = \frac{1}{2}QC$. Straight lines are drawn through P and Q parallel to the sides of the rectangle intersecting at O. Points are taken at equal intervals along the straight line AP and straight lines are drawn from these points through O terminating on QC. Find the mean value of the lengths of these straight lines.

NOTES AND HINTS

FOR SOLUTION OF THE QUESTIONS

4. It is unnecessary to obtain the partial fractions of the expression. Since $4x + 17 = 2(2x+7) + 3$, the given fraction can be separated into two inverse factorials, with interval of differencing 2.

6. Since all the differences of $x^{(m)}$ higher than the mth are zero, the sum of the series to m terms is the same as the sum to infinity.

17. $f(x)$ is zero when x has the values 1, 2 or 3. The curve $y = f(x)$ is of the fifth degree in x. Assume therefore that
$$y = (x-1)(x-2)(x-3) F(x),$$
where $F(x)$ is of the form $ax^2 + bx + c$. The values of a, b and c can then be found from the values of $f(x)$ when x is 0, 4 and 5.

25. This example can, of course, be done by the use of divided differences. It is simpler, however, to change the origin. The required form of function is then self-evident.

33. If advancing differences are used, each of the three calculations necessitates a fresh set of leading differences and a fresh set of coefficients. By using a central difference formula, however, the labour is considerably lessened. The basic formula is unchanged, further terms being added to allow of the inclusion of more extensive data. By this means successive approximations to the required result can be obtained. The example illustrates, in fact, one of the principal advantages that a central difference formula possesses over an advancing difference formula.

39. In using Everett's formula to obtain a succession of interpolated values it is essential that the work be set out in the tabular form given on p. 79 of the Text-book. If u_x is expanded separately for each of the required values of x, the whole advantage of the formula is lost. In this connection it is useful to remember the coefficients of the terms in that part of the formula involving x (as distinct from the part involving ξ): these coefficients are given on p. 78 of the book.

43. As a practical measure, it is a saving of labour if the percentage increase in the pension be found for each age of wife given in the table. In order to obtain the required result one interpolation is all that is necessary instead of the two that would seem to be indicated by the question.

44. Where data are given for regular intervals of n and it is required to find all the values at unit intervals, it is advisable to revert to first principles. The formula adopted should be based on the symbolic identity $(1+\delta)^n \equiv (1+\Delta)$, where δ represents the operation of differencing for unit intervals and Δ the operation for intervals of n. Everett's formula is unsuitable, as interpolated values throughout the range cannot be obtained by the application of this formula.

52. This example is an interesting illustration of the application of the principle of central differences. By the use of Lagrange's formula the required function can be expressed in terms of the known values of the function among which it occupies a central position. In this question, all that need be done is to express $\dfrac{f(x)}{(x^2-\frac{1}{4})(x^2-\frac{9}{4})}$ in partial fractions, and the expansion follows immediately.

67. It is necessary to distinguish between expressions of the form

$$mx(mx-m)(mx-2m)\dots(mx-rm)$$

and

$$px(px-1)(px-2)\dots(px-r).$$

The first of these is a simple factorial, which may be written as $m^{r+1}x^{(r+1)}$. The second cannot be put into factorial form immediately, since the successive factors comprising the expression do not decrease by the required differences.

72. The general term is of the form $p+qx+r^x+s^x$. By differencing successively the following equations will be obtained:

$$r^x(r-1)^2+s^x(s-1)^2=48 \qquad \dots\dots\text{(i)},$$
$$r^x(r-1)^3+s^x(s-1)^3=132 \qquad \dots\dots\text{(ii)},$$
$$r^x(r-1)^4+s^x(s-1)^4=372 \qquad \dots\dots\text{(iii)},$$
$$r^x(r-1)^5+s^x(s-1)^5=1068 \qquad \dots\dots\text{(iv)}.$$

To solve the equations, take all the terms in s to the right-hand side and divide the terms of each side of equation (i) by the corresponding terms of equation (ii); similarly for (ii), (iii) and (iii), (iv). Proceed in this way with the resulting equations and r will be found.

82. Since $\Delta^3 u_x$ is constant, it may be assumed that u_x is a rational integral function of the form $a + bx + cx^2 + dx^3$. The required result can therefore be found by using the facts given and solving the equations thus obtained.

There is, however, a neater solution. We have immediately that $S_1 = 8$, $S_3 = 34$, $S_6 = 173$ and $S_{10} = 1175$, where S_n is the sum of n terms of the series $u_1 + u_2 + u_3 + \dots$. The value required, u_5, is evidently $S_5 - S_4$. If now we write down the four values of S_n given above in the form required for a difference table and take out the successive divided differences, we shall of necessity stop at third differences of S.

But since u_x is of the form $a + bx + cx^2 + dx^3$, it follows that

$$S_x \ (= \Sigma u_x) \text{ is of the form } A + Bx + Cx^2 + Dx^3 + Ex^4;$$

for the effect of integrating a rational integral function of x of degree n is to produce a rational integral function of degree $n + 1$.

Third differences of S are, therefore, not constant, and to obtain the constant fourth difference we must introduce another term. This term is evidently S_0, which is zero, since the sum of no terms is obviously zero.

By writing down the five values of S_x now available and applying the ordinary divided difference formula, any value of u_x can be found.

87. The series consists of nine terms, and as no further information is given we may assume that u_x is a rational integral function of x of the eighth degree in x. In that event $\Delta^9 u_x$ will be zero.

To obtain the result required express $\Delta^9 u_x$ in the form $(E - 1)^9 u_x$, expand and use the data given.

94. In this question, and in many of those which follow, the

method of separation of symbols can be employed to advantage. For example:

$$c_0 a - c_1 (a-1) + c_2 (a-2) - \ldots + (-1)^n c_n (a-n)$$
$$= c_0 a - c_1 E^{-1} a + c_2 E^{-2} a - \ldots + (-1)^n c_n E^{-n} a$$
$$= [c_0 - c_1 E^{-1} + c_2 E^{-2} - \ldots + (-1)^n c_n E^{-n}] a$$
$$= (1 - E^{-1})^n a = \Delta^n E^{-n} a = \Delta^n (a-n) = 0.$$

95. Note that the variable here is $x^{(-1)}$.

96. On adopting the method outlined for the solution of Qu. 94, the series is seen to reduce to an expression of the form

$$(a + b\Delta + c\Delta^2)(x-n)^2.$$

By the use of the properties of differences of zero, namely that

$$\Delta 0^2 = 1; \quad \Delta^2 0^2 = 2; \quad \Delta^n 0^2 = 0 \text{ if } n \text{ is greater than } 2;$$

the answer can be found with little trouble.

98. Put $x = \tfrac{1}{2}n$, so that the series reads

$$c_0 2^n x^n - c_1 2^n (x-1)^n + c_2 2^n (x-2)^n - \ldots.$$

Find the value of this expression. The result will be found to be independent of x.

101. Consider the more general expression

$$\frac{1}{x} - \frac{1}{x+1}.m + \frac{1}{x+2}.\frac{m(m-1)}{2!} - \frac{1}{x+3}.\frac{m(m-1)(m-2)}{3!} + \ldots.$$

This can be written in the form

$$(1 - Em + E^2 m_2 - E^3 m_3 + \ldots) x^{(-1)}.$$

If this be evaluated and x be put equal to 2 in the result, the sum of the original series will be obtained.

104. Note that the interval of differencing is b and not unity.

106. At first sight this question does not appear to be amenable to the method of treatment advocated for the previous questions. Consider, however, the series

$$\frac{r(r+1)}{2} a_2 + \frac{(r+1)(r+2)}{2} a_3 + \frac{(r+2)(r+3)}{2} a_4 + \ldots.$$

When $r = 1$ this is the series on the left-hand side of the identity that it is desired to prove.

Further, since $\dfrac{(r-2)(r-1)}{2}$ and $\dfrac{(r-1)r}{2}$ are both zero when $r=1$, we may write
$$a_2 + 3a_3 + 6a_4 + 10a_5 + \dots$$
as $\tfrac{1}{2}\{(r-1)^{(2)}a_0 + r^{(2)}a_1 + (r+1)^{(2)}a_2 + (r+2)^{(2)}a_3 + \dots\}$ when $r=1$
$$= \tfrac{1}{2}(a_0 + a_1 E + a_2 E^2 + \dots)(r-1)^{(2)} \quad \text{when } r=1,$$
$$= \tfrac{1}{2}(1 + E + E^2 + \dots + E^{2m})^n (r-1)^{(2)} \quad \text{when } r=1.$$

By summing the progression in the bracket and replacing the operator E by the equivalent $1+\Delta$ the result can readily be obtained.

109. Care must be taken to distinguish between operations on u_x and operations on v_x. Let δ operate on u for differences of $\tfrac{1}{10}$, so that $\delta u_0 = u_{\frac{1}{10}} - u_0$; and let Δ operate on v for unit differences.

For example: $\quad \Delta v_0 = v_1 - v_0$
$$= u_{\frac{10}{10}} - u_0$$
$$= [(1+\delta)^{10} - 1]\,u_0, \text{ etc.}$$

124. Lidstone's method for proving this identity is by the use of Bessel's formula. The substance of his method is as follows:

Let $U_x = u_{x-1} + u_{x-2} + u_{x-3} + \dots$; then $\delta U_x = u_x$, $\delta^2 U_x = \delta u_x$ and so on; and
$$\sum_0^{n-1} u_x = U_n - U_0 = U_{(c+\frac{1}{2})+\frac{1}{2}n} - U_{(c+\frac{1}{2})-\frac{1}{2}n}.$$

Bessel's formula may be written in the form
$$U_{c+\frac{1}{2}+y} = \overline{U} + y\delta U_c + \frac{y^2 - (\frac{1}{2})^2}{2!}\delta^2 U_{c-1} + \frac{y\{y^2 - (\frac{1}{2})^2\}}{3!}\delta^3 U_{c-1} + \dots,$$
where $\overline{U}$, $\delta^2 U$, $\delta^4 U$, ... represent the mean U and the mean even central differences lying on the line of $U_{c+\frac{1}{2}}$. Putting $y = \frac{1}{2}n$ and $-\frac{1}{2}n$ successively and taking the second result from the first, the odd terms cancel because their coefficients are even functions of y; and the even terms are doubled because their coefficients are odd functions of y. We then have
$$[n]\,u_c = U_n - U_0 = n\delta U_c + \frac{2 \cdot \frac{1}{2}n\{(\frac{1}{2}n)^2 - (\frac{1}{2})^2\}}{3!}\delta^3 U_{c-1} + \dots,$$
from which the required formula can easily be obtained. (See *J.I.A.* vol. LV, p. 178.)

An alternative method is to put $\delta \equiv E^{\frac{1}{2}} - E^{-\frac{1}{2}} \equiv e^{\frac{1}{2}D} - e^{-\frac{1}{2}D}$; since $[n] \equiv (E^{\frac{1}{2}n} - E^{-\frac{1}{2}n})/(E^{\frac{1}{2}} - E^{-\frac{1}{2}})$, we can expand E in terms of D and thence in terms of δ.

A third method, which applies only to the case when n is odd, is due to A. W. Sunderland.

By equating coefficients in the identity
$$\log (1 - px) + \log (1 - qx) = \log (1 - \overline{p + q}x + pqx^2)$$
and putting $1 + \Delta$ for p and -1 for q, an expression for $(1 + \Delta)^n - 1$ is obtained; whence by dividing by Δ a series of operators for $\overset{n-1}{\underset{0}{\Sigma}}$ is at once evident. (See *J.I.A.* vol. LV, p. 179.)

126. The student is advised to read the note by Professor Steffensen "On the Definition of the Central Factorial" in *J.I.A.* vol. LXIV, pp. 165 *et seq.*, where well-known finite difference formulae can at once be obtained from the general formulae given therein.

127. This example and the two following ones are simple exercises on linear difference equations. A very lucid explanation of the elementary principles of this subject by P. M. Marples will be found in *J.I.A.* vol. LXIII, pp. 404-423. The illustrative examples on the last few pages of this note are of particular interest.

135. This is essentially a question which is most easily solved by logarithmic differentiation.

Take logarithms of both sides:

then
$$\frac{m}{m+n} \log x + \frac{n}{m+n} \log y = \log (x + y).$$

Differentiate with respect to x:
$$\frac{m}{m+n} \cdot \frac{1}{x} + \frac{n}{m+n} \cdot \frac{1}{y} \cdot \frac{dy}{dx} = \frac{1}{x+y} \left\{ 1 + \frac{dy}{dx} \right\}.$$

Multiply both sides by $m + n$, simplify, and the result follows.

136. $\operatorname{Tan} 2\alpha = 2 \tan \alpha / (1 - \tan^2 \alpha)$, so that if $x = \tan \alpha$,
$$\tan^{-1} \{ 2x / (1 - x^2) \} = \tan^{-1} (\tan 2\alpha) = 2\alpha.$$
Similarly $\sin^{-1} \{ 2x / (1 + x^2) \} = \sin^{-1} (\sin 2\alpha) = 2\alpha$, from the corresponding formula for $\sin 2\alpha$ in terms of $\tan \alpha$.

141. Note that when differentiating a function of the form $\log \dfrac{f(x)}{F(x)}$ it is unnecessary to use the formula for the differentiation of a quotient. Write the expression as $\log f(x) - \log F(x)$ and differentiate each term separately.

149. It is essential to take care when differentiating complicated forms involving indices. For example, in the second part of the question:

$$u = a^{x^n}$$

so that

$$\log u = x^a \log a,$$

whence du/dx follows immediately.

$$v = x^{a^x}$$

and

$$\log v = a^x \log x,$$

in which case

$$\frac{\mathrm{I}}{v} \cdot \frac{dv}{dx} = a^x \log a \log x + \frac{a^x}{x}.$$

From these two results dz/dx can at once be found.

155. The simplest method for the solution of this question is to write

$$y = p + qx + rx^2 + sx^3 + \dots$$

Then

$$dy/dx = q + 2rx + 3sx^2 + \dots,$$

and the successive derivatives when $x = 0$ are q, $2r$, $6s$, ...

The equation of the curve is

$$ax + by + cx^2 + dxy + ey^3 = 0.$$

Therefore

$$ax + b\,(p + qx + rx^2 + sx^3) + c\,(p + qx + rx^2 + sx^3)^2$$
$$+ dx\,(p + qx + rx^2 + sx^3) + e\,(p + qx + rx^2 + sx^3)^3 = 0.$$

Since the curve passes through the origin, $p = 0$, and to obtain q, r and s all that is necessary is to equate the coefficients of x, x^2 and x^3 in the identity above.

158. By differentiating in the ordinary way with respect to x an unwieldy expression is obtained. It is advisable therefore to seek for a simpler relation between y and x than that given by the form $2x = y^2 + \mathrm{I}/y^2$.

Square both sides of the equation.

Then
$$4x^2 = y^4 + 2 + 1/y^4,$$
$$4x^2 - 4 = y^4 - 2 + 1/y^4,$$
$$4(x^2 - 1) = (y^2 - 1/y^2)^2,$$

or, taking the positive sign,
$$y^2 - 1/y^2 = 2(x^2 - 1)^{\frac{1}{2}}.$$

But
$$y^2 + 1/y^2 = 2x,$$
$$\therefore\ y^2 = (x^2 - 1)^{\frac{1}{2}} + x,$$
$$\therefore\ 2yDy = x(x^2 - 1)^{-\frac{1}{2}} + 1 = \{x + (x^2 - 1)^{\frac{1}{2}}\}/(x^2 - 1)^{\frac{1}{2}}$$
$$= y^2/(x^2 - 1)^{\frac{1}{2}},$$
$$\therefore\ y = 2(x^2 - 1)^{\frac{1}{2}} Dy,$$

since y is not zero.

Differentiate again and the required result is easily found.

159. A useful principle which can be applied to many questions of this type (and in fact to many other questions in differential calculus) is that it is easier to differentiate a product than to differentiate a quotient.

Thus
$$y = (1 - x^2)^{\frac{1}{2}} \sin^{-1} x,$$

so that
$$\frac{dy}{dx} = \frac{(1 - x^2)^{\frac{1}{2}}}{(1 - x^2)^{\frac{1}{2}}} - \frac{x \sin^{-1} x}{(1 - x^2)^{\frac{1}{2}}} = 1 - \frac{x \sin^{-1} x}{(1 - x^2)^{\frac{1}{2}}}.$$

If now instead of differentiating this relation as it stands, we clear of fractions,
$$(1 - x^2)^{\frac{1}{2}} dy/dx = (1 - x^2)^{\frac{1}{2}} - x \sin^{-1} x,$$

and then replace $\sin^{-1} x$ by $y(1 - x^2)^{-\frac{1}{2}}$, we have, after multiplication by $(1 - x^2)^{\frac{1}{2}}$,
$$(1 - x^2) dy/dx = (1 - x^2) - xy.$$

This is a convenient form from which to obtain the relation required.

174. If $u = (x^2 + 1)^{\frac{1}{2}} \log\{x + (x^2 + 1)^{\frac{1}{2}}\}$, it is easy to show that
$$(x^2 + 1) Du = x^2 + 1 + xu.$$

Let
$$u = a_0 + a_1 x + a_2 x^2 + \ldots + a_n x^n + \ldots.$$

Then
$$Du = a_1 + 2a_2 x + \ldots + na_n x^{n-1} + \ldots.$$

In the relation $(x^2 + 1) Du = x^2 + 1 + xu$ replace u and Du by these series, equate coefficients of the various powers of x and $a_1, a_2, \ldots, a_r, \ldots, a_n$ will be found. a_0 is obviously 0, since u is 0 when x is 0.

181. Since four facts about u_x are given, we must assume that u_x is a rational integral function of the third degree in x. We may therefore write

$$u_x = (x-2)(ax^2+bx+c),$$

since u_2 is o.

a, b and c can then be found by using the information given.

191. It is unnecessary, and very cumbersome, to attempt to use the methods of the calculus in solving this problem. A careful drawing should be made embodying the particulars given, and the result obtained by reference to the geometry of the figure.

199. Although the differential coefficients of y with respect to x are required, it is of advantage to maintain α as the variable.

$$y = \alpha/\sin\alpha \quad \text{and} \quad x = 1 - \cos\alpha.$$

$$\frac{dy}{dx} = \frac{dy}{d\alpha}\Big/\frac{dx}{d\alpha} = \frac{\sin\alpha - \alpha\cos\alpha}{\sin^3\alpha}.$$

Similarly $\quad \dfrac{d^2y}{dx^2} = \dfrac{\alpha\sin^2\alpha - 3\sin\alpha\cos\alpha + 3\alpha\cos^2\alpha}{\sin^5\alpha}.$

204. A straightforward method for the proof of the identity is as follows:

Denote differentiations with respect to θ as under:

$$x' = dx/d\theta; \quad x'' = d^2x/d\theta^2; \quad y' = dy/d\theta \quad \text{and so on.}$$

Then, from the first equation,

$$(x-a)^2 + (y-b)^2 = c^2,$$

$$\therefore \quad (x-a)x' + (y-b)y' = 0$$

and $\quad (x-a)x'' + (y-b)y'' = -\{(x')^2 + (y')^2\}.$

Also, since $\quad x = r\cos\theta \quad \text{and} \quad y = r\sin\theta,$

$$x' = r'\cos\theta - r\sin\theta$$

and $\quad y' = r'\sin\theta + r\cos\theta,$

so that $\quad (x')^2 + (y')^2 = r^2 + (r')^2.$

Again $\quad x'' = r''\cos\theta - 2r'\sin\theta - r\cos\theta$

and $\quad y'' = r''\sin\theta + 2r'\cos\theta - r\sin\theta.$

Eliminate $x-a$ and $y-b$ from the first three equations above, and substitute for the differential coefficients of x and y to obtain the result.

205. Although the result can be obtained by immediate differentiation of both sides of the given equations, it is simpler to solve the equations for x^2 and y^2.

We find, quite easily, that $x^2 = (c+\alpha)(c-\beta)/c$ and $y^2 = \alpha\beta/c$; whence

$$\alpha - \beta = x^2 + y^2 - c \quad \text{and} \quad \alpha\beta = cy^2.$$

Differentiate these equations partially with respect to x and y and, after a few straightforward substitutions, the result becomes evident.

206. The relationship between D and Δ given on p. 213 of the Text-book is

$$D \equiv \log(1+\Delta).$$

This holds only when the interval of differencing is unity. If the interval is h, then

$$u_{x+nh} = E^n u_x.$$

Also

$$u_{x+nh} = e^{nhD} u_x$$

by Taylor's theorem.

Therefore

$$E^n \equiv e^{nhD}$$

or

$$hD \equiv \log(1+\Delta)$$

and

$$D \equiv \frac{1}{h} \log(1+\Delta)$$

$$\equiv \frac{1}{h} \left\{ \Delta - \frac{\Delta^2}{2} + \frac{\Delta^3}{3} - \dots \right\}.$$

In this question $h = 5$, so that, after taking out the successive differences of the function, we must write

$$\left[\frac{dl_x}{dx} \right]_{x=10} = \frac{1}{5} \left[\Delta - \frac{\Delta^2}{2} + \frac{\Delta^3}{3} - \dots \right] l_{10}.$$

Similar considerations apply to Qu. 210 and Qu. 212.

211. By Lagrange's formula of interpolation,

$$u_x = \frac{(x-b)(x-c)(x-d)}{(a-b)(a-c)(a-d)} u_a + \text{three similar expressions}.$$

Differentiate with respect to x, put $x = 0$, and the answer to the first part of the question is at once obtained.

For the second part of the question, put $u_x = x^3$ in the result and clear of fractions.

F E

213. It is easy to show that, if $\Delta x = 1$, $\Delta u_x = (x+1)\log(x+1) - x\log x$. Iv_x, which is $\int v_x dx$, will only be $(x+1)\log(x+1) - x\log x$ provided that the constant of integration is zero.

217. The important point to note here is that the presence of the constant of integration is essential. Proceeding in the usual way it is found that

$$\log l_x = 10^{-Ax^{\frac{1}{2}}-B} + \text{a constant}.$$

As a result l_x will be $ke^{10^{-Ax^{\frac{1}{2}}-B}}$, where k is independent of x.

227. The results obtained by the methods suggested in this question and in the following one would seem to be different, although the same integral is involved. It is instructive to express the answers to the two questions in terms of x and to see wherein they differ.

233. The answer to the second integral is $\sin^{-1}\frac{1}{2}$. The value of a definite integral involving such an expression is usually the smallest positive angle: the result is therefore $\pi/6$. (See Text-book, p. 265, Note (2).)

236. Proceeding in the usual way, the value of the integral is found to be -2. This is obviously impossible, for every term in $h\Sigma f(x)$—where $f(x)$ is $(1-x)^{-2}$—is positive. The limit cannot therefore be negative. Further, the integrals $\int_0^1 \dfrac{dx}{(1-x)^2}$ and $\int_1^2 \dfrac{dx}{(1-x)^2}$ are both infinite.

The example shows therefore that the ordinary rules for integrating between given limits cannot be adopted when the function to be integrated becomes infinite between those limits.

240. The general term in the series is $\dfrac{1}{\sqrt{r}(\sqrt{n}+\sqrt{r})}$. This can be expressed as $\dfrac{1}{n}\dfrac{1}{\sqrt{r/n}(1+\sqrt{r/n})}$, where r increases by increments of unity from n to $4n$.

The limit of the series is therefore $\int_1^4 \dfrac{dx}{\sqrt{x}(1+\sqrt{x})}$.

246. In the second part of the question the difficulty is to find a suitable substitution which will render the expression integrable. Of the two factors $x^{\frac{1}{2}}$ and $(x^3+1)^{\frac{1}{2}}$, the more suitable one to choose for the purpose of a substitution would seem to be $x^{\frac{1}{2}}$. In order to obtain $x^{\frac{1}{2}}dx$ the function to be used is $x^{\frac{3}{2}}$. Put therefore $x^{\frac{3}{2}}=y$. Then $dx^{\frac{3}{2}}=\frac{3}{2}x^{\frac{1}{2}}dx$, and $(x^3+1)^{\frac{1}{2}}=(y^2+1)^{\frac{1}{2}}$.

The integral is therefore $\int \frac{2}{3}(y^2+1)^{\frac{1}{2}}\,dy$, which can be evaluated at once.

262. To evaluate $\int e^{4x}x^2\,dx$ put $x=\log y$. The integral then becomes

$$\int y^3\,(\log y)^2\,dy,$$

which is, of course, the same as the previous integral in the question.

264. There are several methods by means of which the integral of $(x+1)^{-2}(x^2+1)^{-\frac{1}{2}}$ can be found. A simple substitution is $y=(x+1)^{-1}$. The integral then becomes

$$\int_0^1 \frac{y\,dy}{\sqrt{(1-2y+2y^2)}}.$$

This may be written as

$$\frac{1}{\sqrt{2}}\int_0^1 \frac{(y-\frac{1}{2})\,dy}{\sqrt{(y-\frac{1}{2})^2+\frac{1}{4}}}+\frac{1}{2\sqrt{2}}\int_0^1 \frac{dy}{\sqrt{(y-\frac{1}{2})^2+\frac{1}{4}}},$$

each part of which is of standard form.

270. We require here $\int_0^1 \frac{x^{\frac{3}{2}}dx}{(1-x)^{\frac{1}{2}}}$. If x be put equal to $\sin^2\alpha$, the integral is transformed into $2\int_0^{\frac{\pi}{2}} \sin^4\alpha\,d\alpha$.

By expressing $\sin^4\alpha$ in terms of multiple angles, thus,

$$\sin^4\alpha=\frac{3}{8}-\frac{1}{2}\cos 2\alpha+\frac{1}{8}\cos 4\alpha,$$

the integral can easily be evaluated.

Note that the area of the whole part included between the curve and the asymptote is twice the value of the integral given above.

278. This is perhaps a simpler method for the development of Weddle's rule than that given on pp. 294 and 295 of the Text-book. The method suggested by this question possesses the additional advantage that it eliminates the principal error terms in each of the two formulae employed.

283. The formula here given is simply Lubbock's formula with the addition of the term u_r.

If third and higher differences are zero, we may write

$$\Delta u_r = \Delta^2 u_{r-2} + \Delta u_{r-2}; \quad \Delta^2 u_r = \Delta^2 u_{r-2}.$$

These do not involve a knowledge of the values of u_r when x is greater than r.

286. Although this example is not an example in integral calculus, it has been inserted here in order to show the similarity between the two formulae

$$\int_{a-h}^{a+h} u_x dx = \tfrac{1}{3}h\left(u_{a-h} + 4u_a + u_{a+h}\right)$$

and

$$u_{a+h} - u_{a-h} = \tfrac{1}{3}h\left(u'_{a-h} + 4u'_a + u'_{a+h}\right).$$

The first formula gives the area of the curve $y = u_x$ between the ordinates $x = a-h$ and $x = a+h$, with an error of the order $\Delta^4 u$; while the second gives the difference between the lengths of the ordinates $x = a-h$ and $x = a+h$, with an error of the order $D^5 u$.

287. The true value of the integral is $2 \cdot 14286\dots$, and the approximate value is $2 \cdot 189\dots$ The function $y = (1+0 \cdot 3x)^{-2}$ is changing rapidly over the interval $x = 0$ to $x = 6$, and the approximate integration formula does not therefore give a satisfactory value for the integral. (See Text-book, p. 299, Ex. 3.)

292. The Euler-Maclaurin formula is

$$\frac{1}{r}\int_a^{a+nr} u_x dx = \sum_{x=a}^{x=a+(n-1)r} u_x + \tfrac{1}{2}\left(u_{a+nr} - u_a\right) - \frac{r}{12}\left(u'_{a+nr} - u'_a\right)$$

so that

$$\frac{1}{r}\int_a^{a+nr} u_x dx + \frac{r}{12}\left(u'_{a+nr} - u'_a\right) = \sum_{x=a}^{x=a+(n-1)r} u_x + \tfrac{1}{2}\left(u_{a+nr} - u_a\right).$$

NOTES AND HINTS
69

Put $a=1$, $n=4$, $r=0\cdot1$; then

$$\frac{1}{0\cdot1}\int_1^{1\cdot4} u_x dx + \frac{0\cdot1}{12}(u'_{1\cdot4}-u'_1) = \sum_{x=1}^{x=1\cdot4} u_x + \tfrac12(u_{1\cdot4}+u_1),$$

whence the required result can easily be obtained.

294. By the unitary definition C is right; for the definition gives a measure of the probability by taking into account the actual number of favourable ways out of the total number of ways. C *knows* that there were 9 white balls in the bag; A *thinks* that there were 10. A's answer is therefore wrong. B has no information *a priori*, and is justified in taking the result derived from the experiment as the best available value.

(For further points arising from a consideration of this question, see *J.S.S.* vol. IV, No. 3, pp. 192, 193.)

301. The number of ways in which m different articles can be distributed to n different persons with no restriction as to how many any one person may receive is n^m.

The total number of cases when $m=4$ and $n=4$ is therefore 4^4; the total number when $m=4$ and $n=5$ is 5^4.

The favourable cases can easily be obtained by setting out in detail the various ways of distributing the articles under the conditions laid down by the question.

303. Whichever of the two methods be adopted, the number of ways of making up a total of 24 is the same, namely 9, 9, 6 (3 ways); 9, 8, 7 (6 ways); 8, 8, 8 (1 way).

The total number of ways in which numbers can be drawn is, however, different. If there are three boxes and a man draws a counter from one of the three boxes, he is restricted to the eighteen counters in the second and third boxes for his second draw. If, however, all the counters are in one box and he draws a counter, he has the choice of twenty-six counters for his second draw. The total number of ways of drawing the three counters is therefore not the same in the second case as in the first.

324. If $w=\tfrac12$, $d=\tfrac14$, $l=\tfrac14$, the terms of $(w+d+l)^5$ give the probabilities of all combinations of wins, draws and losses in five games.

The favourable cases are A 3 points and B 2 points; A $2\tfrac12$ points and B $2\frac{1}{2}$ points; A 2 points and B 3 points. The probability that A scores 3 points comes from the terms w^3l^2, w^2d^2l and wd^4; $2\frac{1}{2}$ points from w^2dl^2, wd^3l, d^5; 2 points from w^2l^3, wd^2l^2, d^4l.

The term involving w^3l^2 is $\dfrac{5!}{3!\,2!}\,w^3l^2$ or $\dfrac{5!}{3!\,2!}(\frac{1}{2})^3(\frac{1}{4})^2$; and so on for the other terms.

326. The number of ways of obtaining a 5, 4; 3, 1 distribution is found by multiplying the following expressions together:

The number of ways in which A can have 5 hearts out of 13 hearts;

The number of ways in which A can have 8 other cards;

The number of ways in which B can have 4 hearts out of the remaining 8;

The number of ways in which B can have 9 other cards out of the remaining $(39-8)$ cards;

The number of ways in which Y or Z can have 3 hearts out of the remaining 4;

The number of ways in which Y or Z can have 10 other cards out of the remaining $(26-4)$ cards.

A further factor must be introduced to allow for Y having 3 hearts and Z 1 heart, and *vice versa*.

329. The simplest method of attack is as follows:

If the sequence of three heads commence with the first throw, the result of the other throws is immaterial.

Let H stand for head, T for tail, X for either head or tail.

The chance that the first three throws are heads—i.e. that the sequence is $HHHXXXX$—is $(\frac{1}{2})^3$.

Next, we may have a tail thrown at the start, and then a sequence of three heads, thus: $THHHXXX$. The chance of this event is $(\frac{1}{2})^4$.

This sequence may however occur with a tail thrown at the second, third or fourth throw. The total chance under this heading is therefore $4\,(\frac{1}{2})^4$.

But we have included the sequence $HHHTHHH$ twice, once as $HHHXXXX$ and once as $XXXTHHH$; the chance of this event must therefore be deducted from the sum of the two chances already found.

The required chance is therefore $(\frac{1}{2})^3 + 4(\frac{1}{2})^4 - (\frac{1}{2})^7$.

331. The straightforward method of solving this problem is first to calculate $f(n)$, the probability of success at the nth draw, and then to find the value of $\Sigma n f(n)$ between the limits $n = 2$ and $n \to \infty$.

A more elegant solution is however as under:

The average number of throws necessary to secure a white ball is the sum of the infinite series

$$1 + (\tfrac{8}{10}) + (\tfrac{8}{10})^2 + (\tfrac{8}{10})^3 + \ldots$$

There are now left 8 black balls and 1 white ball, and the average further number of throws required is the sum of the infinite series

$$1 + (\tfrac{8}{9}) + (\tfrac{8}{9})^2 + (\tfrac{8}{9})^3 + \ldots$$

The average number of throws necessary to draw both white balls is the sum of these two results.

332. The steps required for solving this problem are

(i) the chance that all the men will have enlisted in the same battalion;

(ii) the chance that all the men are not in the same battalion;

(iii) the chance that two battalions are in the same division;

(iv) the chance that the men are not all in the same battalion but that the two battalions are in the same division.

The total chance that the conditions of the question are satisfied is (i) + (iv).

The fact that a battalion consists of 1000 men is not really necessary for the solution of the problem, except in so far as 1000 is greater than 100, so that the 100 men can be in one battalion.

336. There are two possible methods here:

(i) The chance required is

$$\frac{1}{2^2} + \left(1 - \frac{1}{2^2}\right)\frac{1}{3^2} + \left(1 - \frac{1}{2^2}\right)\left(1 - \frac{1}{3^2}\right)\frac{1}{4^2} + \ldots \text{ to infinity.}$$

This is the chance that he wins at the first draw; fails to win at the first draw and wins at the second; fails to win at the first and second draw and wins at the third; and so on.

(ii) The chance of failure at the end of n draws is

$$\left(1-\frac{1}{2^2}\right)\left(1-\frac{1}{3^2}\right)\left(1-\frac{1}{4^2}\right)\cdots\left(1-\frac{1}{n^2}\right).$$

The ultimate chance of success is therefore

$$1-\operatorname*{Lt}_{n\to\infty}\left(1-\frac{1}{2^2}\right)\left(1-\frac{1}{3^2}\right)\left(1-\frac{1}{4^2}\right)\cdots\left(1-\frac{1}{n^2}\right).$$

These two results are easily seen to be the same.

337. It is instructive to obtain the solution of this problem by setting out the various steps in detail.

m	Chance of m consecutive heads	Chance of m consecutive heads or tails
20	$\dfrac{1}{2^{20}}$	$\dfrac{4}{2^{21}}$
19	$\dfrac{2}{2^{20}}$	$\dfrac{4}{2^{20}}$
18	$\dfrac{2}{2^{19}}+\dfrac{1}{2^{20}}$	$\dfrac{5}{2^{19}}$
17	$\dfrac{2}{2^{18}}+\dfrac{2}{2^{19}}$	$\dfrac{6}{2^{18}}$
$\vdots$		
10	$\dfrac{2}{2^{11}}+\dfrac{9}{2^{12}}$	$\dfrac{13}{2^{11}}-\dfrac{2}{2^{20}}$

Expectation

$$=\frac{1}{2^{21}}\left(2^{20}-1\right)+\frac{3}{2^{21}}\left(2^{20}-1\right)+\frac{4}{2^{20}}\left(2^{19}-1\right)+\cdots$$

$$+\frac{13}{2^{11}}\left(2^{10}-1\right)-\frac{1}{2^{19}}\left(2^{10}-1\right)$$

$$=\frac{1}{2}-\frac{1}{2^{21}}-\frac{1}{2^9}+\frac{1}{2^{19}}+\sum_{x=10}^{x=20}\frac{2^x-1}{2^{x+1}}(23-x),$$

which can easily be evaluated by finite difference methods.

342. Any positive whole number can be expressed in one of the forms

$$5n,\quad 5n-1,\quad 5n+1,\quad 5n-2\quad\text{or}\quad 5n+2.$$

For a multiple of 5 to be drawn, the numbers must be of the form:

on one domino $5n$ and on the other $5n$,

or   ,, $5n-1$,, $5n+1$,

or   ,, $5n-2$,, $5n+2$.

There are, for example, 5 ways of obtaining a number of the form $5n-1$ on one domino and 6 ways of obtaining a number of the form $5n+1$ on the other.

The chance under this head is therefore $2 \cdot \frac{5}{28} \cdot \frac{6}{27}$; and similarly for the other cases.

361. The chance that X throws 6 with one die is $\frac{1}{6}$ and that Y scores 6 with two dice is $\frac{5}{36}$.

Let u_n be X's chance of winning the match when he has scored n points more than Y.

His chance of winning the next game is $\frac{1}{6} \cdot \frac{31}{36}$, and his chance of winning the match is then u_{n+1}. Similarly, his chance of losing the next game is $\frac{5}{6} \cdot \frac{5}{36}$ and his chance of ultimate success is u_{n-1}.

Again, if X scores 6 and Y scores 6, or if X fails to score 6 and Y also fails to score 6, the chance of X's winning the match remains at u_n:

$$u_n = \tfrac{1}{6} \cdot \tfrac{31}{36} u_{n+1} + \tfrac{5}{6} \cdot \tfrac{5}{36} u_{n-1} + \left(\tfrac{1}{6} \cdot \tfrac{5}{36} + \tfrac{5}{6} \cdot \tfrac{31}{36}\right) u_n.$$

u_0, the chance at the outset, can then be found by either of the methods given in the Text-book (pp. 334, 335).

367. This problem can be solved in any one of a number of ways. Two of the methods are outlined below:

(i) The arrangement when every ticket bought by two half-pennies is bought as soon as possible is

$$P\,P\,2H; \quad P\,P\,2H; \quad P\,P\,2H; \quad P\,P\,2H.$$

Let x_1, x_2, x_3, x_4 be the positions of the four tickets bought by halfpennies.

The condition required is that the arrangements shall be such that

$$x_1 \geqslant 3; \quad x_2 \geqslant 6; \quad x_3 \geqslant 9; \quad x_4 = 12;$$

where

$$x_1 < x_2 < x_3 < x_4.$$

If $x_4 = 12$, $x_3 = 9$, $x_2 = 6$, there are 3 arrangements ($x_1 = 3$, 4 or 5).

If $x_4 = 12$, $x_3 = 9$, $x_2 = 7$, ,, 4 ,, ($x_1 = 3, 4, 5, 6$).

If $x_4 = 12$, $x_3 = 9$, $x_2 = 8$, ,, 5 ,, ($x_1 = 3, 4, 5, 6, 7$).

Thus, if $x_4 = 12$ and $x_3 = 9$, there are in all 12 arrangements.

Similarly, if $x_4 = 12$ and $x_3 = 10$, there are 18 arrangements; and so on.

Since the total number of possible arrangements is $\dfrac{12!}{8!\,4!}$, the required chance is easily obtained.

(ii)

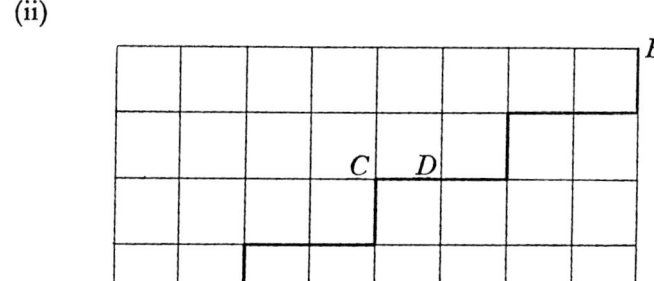

In the diagram each side of a square represents the purchase of a ticket, horizontal movements representing the purchase of a ticket bought by a penny and vertical movements the purchase of a ticket bought by two halfpennies.

The total number of different routes from A to B is the number of ways of arranging 8 things of one kind and 4 of another, namely $\dfrac{12!}{8!\,4!}$.

To satisfy the conditions of the problem the route must not lie above the thick line. The number of favourable routes can easily be counted by inserting the number of ways of reaching the corner of any square, starting from A. Thus, there are evidently 3 ways of reaching C from A, and 7 of reaching D.

371. The chance that a person will not catch cold in a year $= 1 - \cdot926 = \cdot074$.

Let a person catch cold on the average once in t years.

Then the chance of catching cold at any moment $= 1/nt$, where there are n "moments" in a year.

The chance of not catching cold at any moment $= 1 - 1/nt$, and the chance of not catching cold in n moments $= (1 - 1/nt)^n$.

The chance of not catching cold in a year is therefore the limit of the expression $(1 - 1/nt)^n$ as n tends to infinity.

But
$$\operatorname*{Lt}_{n \to \infty} (1 - 1/nt)^n = e^{-1/t}.$$
$$\therefore \quad e^{-1/t} = \cdot074,$$
whence
$$t = \cdot384.$$

$\cdot384$ of a year is about 140 days. Therefore, according to the data given by the question, a person catches cold on the average once in every 140 days.

376. The point to note in this question is that when finding the total number of different ways of printing the pages regard must be had to the position of the pages in space. Thus, the total number of ways of arranging 8 numbers is 8! By turning the paper back to front half the arrangements are repeated, and by rotating the paper through two right angles half the arrangements are again repeated.

The total number of different ways of printing the pages is therefore $\dfrac{8!}{4}$.

379. The mean value of a continuous function $f(x)$ varies according to the law assumed for the successive values of x. (See Text-book, pp. 354, 355.)

In this question an erroneous result will be obtained by the use of the wrong variable. Consider the diagram below.

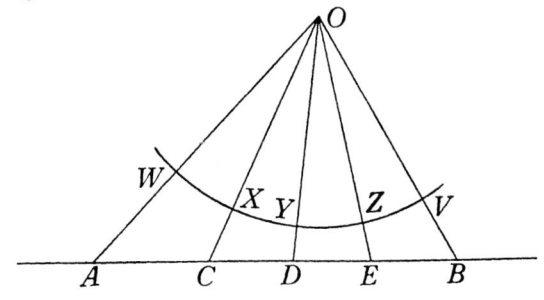

Let O be the position of the object and A and B the two extreme positions of the ship; and let the angles WOX, XOY, YOZ, ZOV be equal. Then the use of the angle at the vertex O of the triangle as the variable implies that this angle changes uniformly during the progress of the ship. If the ship were sailing along the arc $WXYZV$ of the circle with O as centre, equal variations of the angle at O would correspond to equal distances along the arc—i.e. to a uniform speed of the ship along WV. The ship is, however, sailing along the straight line AB. If she were to sail along AB so that the angles at O altered at a uniform rate, she would have to travel from A to C in the same time as from C to D, D to E and E to B. In that event the ship would not be travelling at a uniform speed along AB.

If the mean value is looked upon as the limit of the quotient of the total lengths of all lines drawn from O to AB by the total number of such lines, when the number is indefinitely increased, it is obvious that the result obtained when the lines are drawn so as to divide the base AB into equal parts will not be the same as that obtained when they are drawn so as to divide the angle AOB into equal parts.

380. It is unnecessary to resort to calculus in this question.

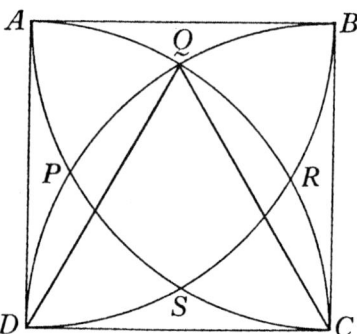

Let $ABCD$ be a square of side a, and let circles of radius a be drawn with A, B, C, D as centres, intersecting at P, Q, R, S.

Then the probability required $= \dfrac{\text{area of curvilinear figure } PQRS}{\text{area of square}}$.

Let $x=$ area of the figure APQ, and $y=$ area of the figure AQB.
Then $y=$ square $ABCD-$ sector $ADQ-$ sector $BCQ-$ triangle
$DCQ=a^2\left(1-\dfrac{\pi}{6}-\dfrac{\sqrt{3}}{4}\right).$

Again, the area of the curvilinear figure $PQRS$ can be expressed as either

(i) square $ABCD-4(x+y)$, or (ii) quadrant $ACD-(3x+2y)$.

Therefore, square $ABCD-4(x+y)=$ quadrant $ACD-(3x+2y)$
$=$ figure $PQRS$.

To obtain the area of $PQRS$ eliminate x from these equations and substitute for the value of y.

384. It is instructive to compare this question with the following one:

A straight line is divided at random into three parts. Find the chance that the three parts of the line can be put together so as to form a triangle. (This is Ex. 3 on p. 360 of the Text-book, worded differently.)

At first sight Qu. 384 and the question above would seem to be the same. There is, however, a real difference.

Consider the case when the line is divided at random into three parts:

$$\overline{\underset{A}{}\underset{P}{}\underset{Q}{}\underset{B}{}}$$
$$\overline{\underset{A}{}\underset{Q}{}\underset{P}{}\underset{B}{}}$$

When the first point P is taken, the second point Q may lie either in PB or in AP; in fact, in the solution given on p. 361 of the Text-book the two cases are considered separately.

When, however, the line is first divided into two parts at P and

$$\overline{\underset{A}{}\underset{Q}{}\underset{P}{}\underset{B}{}}$$

then AP is divided at random into two parts AQ and QP, the second point cannot lie in PB. If the line were actually cut at P, the part PB would not be required further and could be removed, leaving AP to be cut again at Q. In the other question, however, we could not discard PB, as the point Q might fall therein.

To solve Qu. 384, we proceed as follows:

Let $AP = y$ and $AQ = x$. Then for any assigned value of y the chance that Q falls between x and $x + dx$ from A is dx/y. Hence the chance that Q falls between x and $x + dx$ from A and P falls between y and $y + dy$ from A is $\dfrac{dx}{y} \cdot \dfrac{dy}{a}$. To satisfy the conditions of the problem x may vary from $y - \frac{1}{2}$ to $\frac{1}{2}$ and y may vary from $\frac{1}{2}$ to 1.

The chance required is therefore $\displaystyle\int_{\frac{1}{2}}^{1} \int_{v-\frac{1}{2}}^{\frac{1}{2}} \frac{dx}{y} \cdot \frac{dy}{a}$.

390. If

$$M = \mu_y \quad = A + Bc^y,$$
$$N = \mu_{y+1} = A + Bc^{y+1},$$
$$P = \mu_{y+\beta} = A + Bc^{y+\beta},$$

then

$$y + \alpha = y + \frac{N - P}{M - N},$$

so that

$$\alpha = \frac{c^\beta - 1}{c - 1},$$

$$\delta = \beta - \alpha = \beta - \frac{c^\beta - 1}{c - 1}.$$

For (i) differentiate δ with respect to β and proceed in the usual way; for (ii) evaluate $\displaystyle\int_0^1 \delta \, d\beta \Big/ \int_0^1 d\beta$.

398. This is most easily solved by geometry on exactly similar lines to those given in the Text-book. (See p. 367, Method (iii).)

399. It is unnecessary to use integral calculus.

It is easy to see that B is the winner, for the times in which each man completes his course are

A	4 mins. 17·6 secs.
B	4 mins. 16·5 secs.
C	4 mins. 18·4 secs.

B catches up C in 3 minutes, and A catches up C in 4 minutes, so that for 3 minutes the distance between the leading man and the last man is the distance between A and C; for 1 minute it is between A and B; and for the remainder of the time (16·5 secs.) it is between C and B.

By simple arithmetic the average distance required for the whole time is seen to be

$$\frac{(50 \times 180) + (15 \times 60) + (11\tfrac{3}{8} \times 16\tfrac{1}{2})}{256\tfrac{1}{2}} \text{ yards.}$$

ANSWERS TO THE EXAMPLES

1. $225 \, (2^{4x}) - 2$.

2. (i) $6b \left\{ 1 - \dfrac{1}{a+b \, (x+2)^3} \right\}$; (ii) $e^{-2h^2} (e^{hx} - 1) (e^{h^2} - 1)^3$.

3. (i) $\sin \alpha \cos (2x + \alpha)$; (ii) $-\cot h$.

4. $\dfrac{-1920 \, (x+8)}{(2x+1) \, (2x+3) \, (2x+5) \, (2x+7) \, (2x+9) \, (2x+11) \, (2x+13)}$.

5. (i) $\dfrac{4 \, (2x-11)}{(2x-5) \, (2x-3) \, (2x-1)}$; (ii) $\log \dfrac{ax + a^2 + b}{ax - a^2 + b} \log \dfrac{(ax+b)^2 - a^4}{(ax+b)^2}$.

6. $m \, (x-2)^{(m-1)}$.

7. (i) $(x-1) \, e^{x-1} - (x-2) \, e^{x-2}$;

(ii) $9 \left\{ \dfrac{1}{(3x+1) \, (3x+4)} - \dfrac{2}{(3x-1) \, (3x+2)} \right\}$.

8. (i) $e^{x+1} \, (3x^2 + 3x + 1)^{-1} + 6 \, (x-1) - (e^x - 2e^{x-1} + e^{x-2})$; (ii) 0.

10. (i) -117; (ii) $\cdot 1704$.   **11.** -1.

12. $\alpha = 16$, $\beta = 540$.   **13.** $257, 369, 441$.   **15.** $\cdot 02272$.

16. $a_{40} = 13 \cdot 010$, $a_{43} = 11 \cdot 777$.

17. $(x-1) \, (x-2) \, (x-3) \, (x^2 + 8x + 11)$; -47.

18. $u_0 = \cdot 1604$, $u_2 = \cdot 2106$.   **19.** $\cdot 59188$.

20. $2 \cdot 714$, $2 \cdot 758$, $2 \cdot 801$, $2 \cdot 843$, $2 \cdot 884$, $2 \cdot 924$, $2 \cdot 963$, $3 \cdot 001$, $3 \cdot 037$, $3 \cdot 074$, $3 \cdot 109$, $3 \cdot 143$, $3 \cdot 175$.

21. $2 \, (x^3 - 12x^2 + 5x + 48) \, (x+1)^{(-6)}$.   **23.** 967.

24. $x^3 - 2x + 1$.   **25.** $x^3 - 4$.   **26.** $12 \cdot 85$.

28. $\cdot 016018$.   **29.** 363.   **31.** 252 approx.

32. $16 \cdot 25$.   **33.** (i) $22 \cdot 017$; (ii) $22 \cdot 028$; (iii) $22 \cdot 029$.

34. 1191.   **35.** $14 \cdot 6430$.

36. $\cdot 035331$, $\cdot 035335$, $\cdot 035336$, $\cdot 035336 \ldots$ **38.** £14·6061.

39. $\cdot 02031$, $\cdot 02104$, $\cdot 02182$, $\cdot 02264$, $\cdot 02351$, $\cdot 02443$, $\cdot 02539$, $\cdot 02642$, $\cdot 02750$.

40. £2. 10s. 8d., £2. 11s. 11d., £2. 13s. 3d., £2. 14s. 8d., £2. 16s. 2d., £2. 17s. 9d., £2. 19s. 5d., £3. 1s. 2d., £3. 2s. 11d.

41. $\cdot 07094$, $\cdot 07677$, $\cdot 08311$, $\cdot 09000$.

42. $1\cdot4281$, $1\cdot4826$, $1\cdot5399$, $1\cdot6004$, $1\cdot6643$, $1\cdot7320$, $1\cdot8042$, $1\cdot8810$, $1\cdot9629$, $2\cdot0505$, $2\cdot1445$. **43.** $11\cdot6$ per cent.

44. $13\cdot062$.   **45.** $3\cdot262$.   **46.** $1\cdot4502$.

47. $2\cdot07893$, $2\cdot30868$, $2\cdot29376$, $2\cdot29184$, $2\cdot29201$, ...

48. $37\cdot48$, $37\cdot49$.   **49.** $(x-5)^3-2\,(x-5)^2+(x-5)+1$.

52. 1045.

54. $a+(m+n)\,b+\tfrac12 m\,(m-1)\,c+mnc+\tfrac12 n\,(n-1)\,d$, 74.

55. $31\cdot59...$, $29\cdot39...$, or $56\cdot17...$ **56.** $4\cdot688$.

57. (a) $1\cdot2134...$; (b) $2\cdot4695...$

58. $\tfrac{1}{60}(3-p)(29+p)$; 1, 3, -17 or -39.   **59.** $\cdot932$.

60. $\cdot9766$.   **61.** $2\cdot845...$ **62.** (a) $3\cdot091$; (b) $2\cdot917$.

66. (a) $\dfrac{n}{12}(3n^3-22n^2+75n-20)$; (b) $\dfrac{1}{2^{n-3}}+\dfrac{5n^3-27n^2+70n-96}{12}$.

67. $\dfrac{n}{4}(19n^3-54n^2-19n+54)$.

68. (a) $2^{n+2}-4-\dfrac{n}{3}(2n^2-15n+19)$; (b) $2^n-1+\tfrac14 n\,(n-1)\,(n-2)\,(n-3)$.

69. $2^n\{\tfrac12 n\,(n-1)+n+3\}-3$.

70. (a) $\dfrac{29}{36}-\dfrac{6n^2+27n+29}{6\,(n+1)\,(n+2)\,(n+3)}$; (b) $3-\dfrac{6}{(n+1)\,(n+2)}\cdot\dfrac{6^n}{7^n}$.

71. (a) $\tfrac16\,(x^4-18x^3+137x^2-468x+750)$;

(b) $x^2+x+1+2^{7-x}$, $\dfrac{n}{3}(n^2+3n+5)+128-2^{7-n}$.

72. $1+2x+3^x+4^x$, $n^2+2n-\dfrac{17}{6}+\dfrac{3^{n+1}}{2}+\dfrac{4^{n+1}}{3}$.

73. $\dfrac{5n\,(n-1)\,(3n^2+241n-110)}{4}+57+2^{n-1}\,(n^2+57n-114)$.

74. (i) $x.x!$; (ii) $3^x2^{-n}-2^{-x}$.

75. $\dfrac{a^{n+1}}{(a-1)^2}\left\{(n+1)^2-\dfrac{2a\,(2n+3)}{a-1}+\dfrac{6a^2}{(a-1)^2}\right\}$
$\qquad\qquad -\dfrac{a}{(a-1)^2}\left\{1-\dfrac{6a}{a-1}+\dfrac{6a^2}{(a-1)^2}\right\}+nc$.

77. $2^x\,(x^2-3)$, 3072.   **78.** $4-\tfrac12 e$.   **79.** $2.3^{n+1}-5.2^{n+1}+4$.

80. $2^x\,(3x^3-22x^2+71x-113)+c$.

81. $\tfrac{1}{120}(32n+130n^2+85n^3-10n^4+3n^5)$.   **82.** 42.

83. 4.   **84.** $\tfrac16\,(x^3+24x^2+215x+684)$.

F E   6

85. (i) $\frac{1}{20}n(n+1)(n+2)(n+3)(4n+11)$;

    (ii) $\frac{4^{n+2}}{3}(3n^3-3n^2+5n)-\frac{272}{9}(4^n-1)$.   **86.** $14\cdot7,\ 24\cdot6,\ 34\cdot1$.

87. $36\cdot04$.   **88.** $-392\frac{1}{2}$.   **89.** $\frac{1}{3}(5\cdot4^x-2)$.

90. $\frac{1}{9}(-v_{x-1}+5v_x-v_{x+1})$.   **91.** $\cdot0525$.

94. 0.   **95.** $\dfrac{n!}{x(x+1)\dots(x+n)}$.

98. $2^n.n!$.   **99.** 0.   **100.** $2x^2+4x+6$.

101. $\dfrac{1}{m^2+3m+2}$.   **103.** $\dfrac{m}{m-1}$.

104. $\frac{1}{2}(-1)^n b^n (2a+nb)(n+1)!$.   **105.** 0.

111. $\cos\left(\alpha+\dfrac{n-1}{2}\beta\right)\sin\dfrac{n\beta}{2}\operatorname{cosec}\dfrac{\beta}{2}$.   **112.** $\dfrac{\pi}{4}$.

113. 594.   **114.** $1\cdot159$.   **115.** $\cdot487$.

116. $13\cdot147$.   **117.** $33\cdot15$.   **119.** $13\cdot90$.

120. $£2.\ 19s.\ 3d$.   **133.** (i) $-\dfrac{2x(3a^2+x^2)}{3(a^2+x^2)^{\frac{4}{3}}(a^2-x^2)^{\frac{1}{3}}}$;   (ii) $\dfrac{2x^2 e^{\frac{y}{x}}-y}{x(x-y)}$.

135. $\dfrac{y}{x}$.   **136.** 1.   **137.** $\dfrac{-k(ac-b^2)}{(bx+cy)^3}$.

138. (i) $(x-\cos x)^{\sec x}\left[\sec x\tan x\log(x-\cos x)+\dfrac{\sec x(1+\sin x)}{x-\cos x}\right]$;

    (ii) $-ae^{-bx}[b\cos(nx+c)+n\sin(nx+c)]$.

139. (i) $\dfrac{x-1}{x^2(x+1)}+\log\dfrac{x(x+2)}{(x+1)^2}$;   (ii) $\dfrac{e^x}{e-1}\left[x^2-\dfrac{2x}{e-1}+\dfrac{3e-e^2}{(e-1)^2}\right]$.

140. $(0,0),\ (1,0),\ (2,0),\ \left(1-\dfrac{1}{\sqrt{3}},\ \pm\sqrt{\dfrac{2}{3\sqrt{3}}}\right)$.

141. $\dfrac{7}{x^{12}-1}$.

143. (i) $y\cot y\left(\log x\cot x+\dfrac{1}{x}\log\sin x\right)$;   (ii) $\frac{1}{2}x^{-2}\log_{10}e$.

145. $e^{\frac{n}{m}\sin^{-1}\frac{x}{a}}$.

147. $(\log_e a)^2\{x\log(1+x^2)+2x^3(4+2x^2)(1+x^2)^{-2}\}$.

149. (i) $\dfrac{1}{\sqrt{1-x^2}}\left[1+\dfrac{1}{\sqrt{1-u^2}}\right]$;

    (ii) $a^{x^a}ax^{a-1}\log a+x^{a^x}(x^{-1}a^x+a^x\log a\log x)$;

    (iii) $x^{-3}[x-3+(2-x)\log x]$.

152. ·05 inch per second.   **153.** $-3x\,(2y-x)^{-5}$.

155. $-\dfrac{a}{b},\ -\dfrac{2\,(bc-ad)}{b^2},\ \dfrac{6\,(a^3e-abd^2+b^2cd)}{b^4}$.

156. $-3a^2z\,(z^2+a^2)^{-\frac{5}{2}}$.   **160.** $x^2a^{n+2}e^{ax}$.   **161.** n^2.

163. $1+\frac{3}{2}x+\frac{11}{8}x^2+\frac{23}{16}x^3+\frac{179}{128}x^4$, $(2n-3)\,a_{n-2}+3a_{n-1}-2a_n=0$.

165. $3x+\frac{13}{2}x^3+\frac{61}{8}x^5$.   **171.** $x-\frac{2}{3}x^3+\frac{8}{15}x^5$.

172. $\dfrac{x^2}{ab}+\dfrac{x^3\,(a+b)}{a^2b^2},\ \dfrac{x^n n!}{a-b}\left[\dfrac{1}{b^{n-1}}-\dfrac{1}{a^{n-1}}\right]$.

173. $\log_e 2-\frac{1}{8}x^2-\frac{3}{64}x^4-\frac{5}{192}x^6$.   **174.** $x+\dfrac{x^3}{3}-\dfrac{2x^5}{15}$.

175. $1+z+\frac{3}{2}z^2+\frac{8}{3}z^3$ where $z\equiv xe^{-x}$.   **176.** $\log 2-\dfrac{x^2}{4}-\dfrac{x^4}{96}$.

177. $\dfrac{np}{p+q}$ and $\dfrac{nq}{p+q}$.   **178.** $\sqrt{2}+1,\ -\sqrt{2}+1$.   **179.** 2.

180. $-2\frac{2}{3}$.   **181.** 6156.   **183.** 43·36 feet.

185. -8.   **188.** $\dfrac{27c^4}{128a^2}$.   **189.** $\dfrac{3\sqrt{3}}{4}\,w$.

191. The eleventh stair.

195. $-3\sin^2 a\,(a\cos a-\sin a)/a,\ (a^3-\sin^3 a)/\sin a$.

197. e^3.   **198.** -1.   **199.** $\frac{4}{15}$.

202. $4\sin^2\frac{1}{2}x$.   **206.** ·00413.   **207.** $-·0635$.

208. ·903, ·054.   **209.** 34.   **210.** ·0008685, ·43425.

211. 0.   **212.** ·8, ·00048.

214. $\frac{1}{6}x^3-a\log x+x+\frac{1}{6}$.

215. (i) $\frac{2}{3}x^{\frac{3}{2}}\,(1+\frac{1}{3}x^3)$; (ii) $\frac{5}{6}\log(3x^2-10x+10)+\dfrac{22}{3\sqrt{5}}\tan^{-1}\dfrac{\sqrt{5}\,(3x-5)}{5}$.

216. $2\,(a-bc)\,(x+c)^{\frac{1}{2}}+\frac{2}{3}b\,(x+c)^{\frac{3}{2}}$; $\dfrac{1}{(1-x)^{n-1}}\left[\dfrac{1-x}{2-n}-\dfrac{1}{1-n}\right]$.

217. $ke^{10-A\sqrt{x-B}}$.

218. $\frac{1}{3}\log\dfrac{x}{1+x}-\dfrac{1}{3}\dfrac{\log x}{(1+x)^3}+\dfrac{1}{6}\cdot\dfrac{1}{(1+x)^2}+\dfrac{1}{3}\cdot\dfrac{1}{1+x}$; $\dfrac{2a+bx^2}{b^2\sqrt{a+bx^2}}$.

219. $a-\dfrac{1}{4x^4}$.

220. $\frac{1}{2}\{e^x\,(a+b+a\cos x-b\sin x)-b\}+(b-a)\,x$.

221. $x-\frac{1}{2}\log(x+1)-\frac{1}{4}\log(x^2+1)-\frac{1}{2}\tan^{-1}x$,
$\frac{1}{2}(x^2+2ax+2ab-b^2)\log(x+b)-\frac{1}{4}(x^2+4ax-2bx)$.

222. (i) $(x^2-2)\sqrt{1+x^2}$; (ii) $\frac{2}{27}\sin^{-1}\frac{3}{2}x-\dfrac{x}{18}\sqrt{4-9x^2}$; (iii) $\sqrt{1-x^2}/x$.

223. $-[(1+x)+3(1+x)^{\frac{2}{3}}+6(1+x)^{\frac{1}{3}}+6\log\{1-(1+x)^{\frac{1}{3}}\}]$.

224. $\dfrac{1}{4}\left[\sec\theta-\dfrac{1}{2}\log\dfrac{1+\cos\theta}{1-\cos\theta}\right]$,

$\frac{1}{6}\log(1-\cos\theta)+\frac{1}{2}\log(1+\cos\theta)-\frac{2}{3}\log(1+2\cos\theta)$.

225. $\frac{1}{2}\{x\sqrt{x^2-a^2}-a^2\log(x+\sqrt{x^2-a^2})\}$.

226. $\log\{(x^2+a^2)+\sqrt{x^4+3a^2x^2+a^4}\}-\log x$.

227. $\dfrac{1}{\sqrt{7}}\log\dfrac{z-2-\sqrt{7}}{z-2+\sqrt{7}}$, where $z=2x+\sqrt{1+2x+4x^2}$.

228. $-\dfrac{1}{\sqrt{7}}\log[7y+5+\sqrt{7(7y^2+10y+4)}]$, where $y=\dfrac{1}{x-1}$.

229. $\dfrac{1}{55}\dfrac{x^{\frac{5}{3}}(4x^2-11)}{(2x^2-3)^{\frac{11}{10}}}$.   **230.** $A/(c-x)^ne^k$, where $k\equiv\dfrac{1}{a(n-1)(c-x)^{n-1}}$.

231. $\frac{1}{8}e^2+\frac{3}{8}$.   **233.** $\frac{1}{12}\log\frac{5}{2}$, $\dfrac{\pi}{6}$, $\log\frac{3}{2}-\frac{5}{8}$.

234. $a^{m+1}\left[\dfrac{(\log a)^2}{m+1}-\dfrac{2\log a}{(m+1)^2}+\dfrac{2}{(m+1)^3}\right]-\dfrac{2}{(m+1)^3}$.

235. $-\cdot088$.   **237.** $\cdot1622....$ **238.** 686.

240. $2\log_e\frac{3}{2}$.   **244.** $\dfrac{\sqrt{3}}{4}\left[\dfrac{\pi}{3}+\log(2+\sqrt{3})\right]$.

245. $15,\ 23,\ 37,\ 60,\ 95$.

246. (i) $5-6\log_e 2$;   (ii) $\frac{1}{3}\{x^{\frac{3}{2}}\sqrt{1+x^3}+\log(x^{\frac{3}{2}}+\sqrt{1+x^3})\}$.

247. $\cdot67$.   **248.** (i) $\frac{1}{8}$;   (ii) $\frac{1}{12}$.

250. $\dfrac{1}{\sqrt{24}}\log\dfrac{11-2\sqrt{24}}{5}$, $\dfrac{1}{6}$, $\dfrac{1}{\sqrt{6}}\tan^{-1}\dfrac{\sqrt{6}}{7}$.

252. $\dfrac{e^{\frac{a\pi}{2}}[\pi a(1+a^2)+2(1-a^2)]-4a}{2(1+a^2)^2}$.

254. (i) $\frac{1}{2}$;   (ii) $\frac{4}{75}(1+x^{\frac{5}{2}})^{\frac{3}{2}}(3x^{\frac{5}{2}}-2)$.

255. $\frac{1}{2}\sqrt{21}-\sqrt{3}$.   **256.** $ax+b,\ ax^2+2bx+c,\ ax^3+3bx^2+3cx+d$.

258. (i) $\frac{1}{2}x^2+\dfrac{1}{2x^2}+\log\dfrac{x}{1+x^2}$;   (ii) $\frac{16}{105}$.   **260.** $\frac{1}{4}-\frac{1}{4}\log 2$.

262. $\dfrac{x^4}{32}[8(\log x)^2-4\log x+1]$.

264. (i) $\dfrac{1}{2\sqrt{2}}\log(3+2\sqrt{2})$;   (ii) $\dfrac{1}{5}\left[1-\dfrac{e^{\frac{\pi}{8}}}{2\sqrt{2}}\right]$.   **265.** πc^2.

267. 13.   **268.** $\frac{8}{45}\sqrt{3}$.   **269.** $\dfrac{37a^2}{162}$.

271. π.   **272.** $\pi-2$.   **274.** $a^2(2-\frac{1}{2}\pi)$.

276. $a^2\left(1 - \tfrac14\pi\right)$. **277.** $\dfrac{5\sqrt{2}-2}{6}a^2$. **279.** 7·723, 7·535.

280. $\dfrac{2h}{45}[7u_{-2h} + 32u_{-h} + 12u_0 + 32u_h + 7u_{2h}]$. **281.** 3·142.

287. 2·189.... **288.** ·4343.... **289.** 46,801,530.

290. 697 gallons. **292.** −·583.

293. (1) $\tfrac{1}{64}$; (2) $\tfrac{27}{64}$; (3) $\tfrac{27}{64}$; (4) $\tfrac{27}{64}$; (5) $\tfrac{1}{16}$; (6) $\tfrac{3}{8}$; (7) $\tfrac{1}{2917}$; (8) $\tfrac{1728}{2197}$;

(9) $\tfrac18$; (10) $\tfrac34$. **295.** $\tfrac{13}{20}$. **296.** (i) 40; (ii) $\tfrac{3}{20}$.

297. $\tfrac12, \tfrac14$. **298.** $\tfrac{19}{240}$. **299.** $\tfrac{1}{47804}$.

300. $\tfrac{375}{1024}$. **301.** (i) $\tfrac{9}{16}$; (ii) $\tfrac{72}{125}$.

302. Bag No. 1   30 white, 60 red, 10 blue.

,,   2   40 ,,   40 ,,   20 ,,

,,   3   20 ,,   50 ,,   30 ,,

303. $\tfrac{10}{729}, \tfrac{37}{2925}$. **304.** Between 7 and 8 times. **305.** $\tfrac{17}{2250}$.

306. $\tfrac{93}{8192}$. **307.** $\tfrac{323}{2048}$. **309.** $\tfrac{11}{18}$.

310. $\dfrac{5164696}{6^9}$. **312.** $\tfrac12$. **313.** $\tfrac{3}{16}$.

314. 15 (approx.). **315.** $\tfrac{67000}{128578}$. **316.** 7:2.

317. $\dfrac{a^2(a+3c)}{(a+c)^3}$. **318.** ·023. **319.** $\tfrac{13}{2592}$.

320. $\tfrac{22}{25}$. **321.** 4·543. **322.** $\tfrac{1}{30045015}$.

323. (i) $\tfrac{64}{129}, \tfrac{40}{129}, \tfrac{25}{129}$; (ii) $\tfrac{27}{56}, \tfrac{18}{56}, \tfrac{11}{56}$. **324.** $\tfrac{119}{258}$.

325. $\tfrac{1}{78}$. **327.** $\tfrac{13}{36}$. **328.** ·32.

329. $\tfrac{47}{128}$. **330.** $\tfrac{46}{833}$. **331.** 14.

332. $\dfrac{1}{23}\left\{5 + \dfrac{9}{2^{98}}\right\}$. **333.** $\tfrac{333}{2197}$. **334.** $\tfrac{5}{7}$.

335. 3. **336.** $\tfrac12$. **337.** 44 shillings (approx.).

338. $\tfrac{1}{45}\left(\tfrac{19}{40}\right)^{999}$. **339.** Twopence. **341.** 5 hours 20 minutes.

342. $\tfrac{25}{126}$. **343.** (i) $\tfrac89$; (ii) $\tfrac{8}{81}$; (iii) $\tfrac{65}{81}$.

344. $\tfrac35, \tfrac34, 1\tfrac{5}{16}$. **345.** 287/6⁵. **346.** 10 shillings.

347. (i) $\tfrac{11}{60}$; (ii) 5 shillings; (iii) 4s. 11½d. **348.** 13s. 4d.

349. $\tfrac{859}{1344}$. **350.** $\dfrac{1}{r^r}\left[\dfrac{r(r-1)}{2} - \dfrac{(2r-1)!}{\{(r-1)!\}^2} + \dfrac{(3r-1)!}{2r!\,(r-1)!}\right]$.

351. 1s. 2¾d. approx. **352.** $\tfrac59$. **353.** $\tfrac{37}{60}$.

354. $\tfrac{1}{36}, \tfrac{1}{972}$. **355.** $\tfrac{1}{48}$. **356.** $\tfrac{11}{20}$.

357. ·65505.... **358.** $\tfrac12$. **359.** $\tfrac{286}{805}$.

360. $\tfrac{16}{15}\left(1\tfrac{3}{13}\right)^n - \tfrac{1}{15}\left(-\tfrac{3}{13}\right)^n$. **361.** $\dfrac{1-\left(\tfrac{25}{31}\right)^2}{1-\left(\tfrac{25}{31}\right)^{12}}$.

362. $1 - \frac{1}{4}x - \frac{3}{16}x^2 - \frac{9}{64}x^3$, $\frac{67}{256}$.

363. $1\frac{54}{746}$.

364. $\frac{149}{2035}$.

366. $\dfrac{4n-5}{(2n-1)(2n-3)}$.

367. $\frac{1}{9}$.

368. $7d$. approx.

369. $\frac{7}{12}$, $\frac{17}{54}$, $\frac{11}{108}$.

370. $\frac{7}{10}$.

371. Once every 140 days.

372. $\dfrac{(n-3)(n-4)}{n(n-1)}$.

373. $\frac{16}{81}$.

375. $\cdot 3872$.

376. $\frac{1}{630}$.

377. $\dfrac{1}{\sqrt{8}}$.

378. $\frac{9}{32}$.

379. $1 \cdot 57$ miles.

380. $\cdot 315$.

381. $\frac{329}{3245}$.

382. $\frac{7}{10}$.

383. $\frac{4}{25}$.

384. $\log_e 2 - \frac{1}{2}$.

385. $\dfrac{n}{200} - \dfrac{2n^3}{3 \cdot 10^6}$.

386. $\frac{1}{\pi} \cos^{-1}\left(-\frac{1}{15}\right)$.

387. $\dfrac{4a}{\pi}$.

388. $\cdot 020$.

389. $\dfrac{\sqrt{3} + \frac{1}{2}\log(2+\sqrt{3})}{2\sqrt{3}}$.

390. (i) $\dfrac{1}{\log_e c}\left[\log_e \dfrac{c-1}{\log_e c} - 1\right] + \dfrac{1}{c-1}$;   (ii) $\dfrac{1}{2} + \dfrac{1}{c-1} - \dfrac{1}{\log_e c}$.

391. $\left\{\dfrac{\sqrt{3}\pi}{9} - \dfrac{1}{4}\right\} r$.

392. $\dfrac{8r(4-\pi)}{\pi^2}$.

393. $\dfrac{4\sqrt{2r}}{\pi}$, $\dfrac{2\log(\sqrt{2}+1)}{r\pi}$.

394. $\dfrac{\pi r}{4}$, $\dfrac{2r}{\pi}$.

395. $\dfrac{\pi b}{4}$, $\dfrac{2b^2}{3}$.

396. $a^2 \log \frac{1}{2}e$.

397. $1\frac{1}{3}$ miles from Y, $\frac{3}{4}$.

398. $\frac{2}{3}$.

399. $39 \cdot 3$ yards.

400. $b\left\{\frac{1}{4}(12\sqrt{2} - 5\sqrt{5}) + \log_e \dfrac{11 + 5\sqrt{5}}{2\sqrt{2}+2}\right\}$.